CE Marking Handbook

Test and Measurement World Series

CE Marking Handbook
A Practical Approach to Global Safety Certification

David Lohbeck

Newnes

Boston Oxford Johannesburg Melbourne New Delhi Singapore

Newnes is an imprint of Butterworth–Heinemann.

Copyright © 1998 by Butterworth–Heinemann

 A member of the Reed Elsevier group

Recognizing the importance of preserving what has been written, Butterworth–Heinemann prints its books on acid-free paper whenever possible.

 Butterworth–Heinemann supports the efforts of American Forests and the Global ReLeaf program in its campaign for the betterment of trees, forests, and our environment.

Library of Congress Cataloging-in-Publication Data

Lohbeck, David, 1950–
    CE marking handbook : a practical approach to global safety certification /
    David Lohbeck.
      p.   cm. — (Test and measurement world series)
    Includes bibliographical references and index.
    ISBN 0–7506–9819–5 (hardcover : alk. paper)
      1. Standardization—Law and legislation—European Union countries.
2. Product safety—Law and legislation—European Union countries.
3. Machinery—Safety regulations—European Union countries.
4. Standardization—European Union countries—Marks. I. Title.
II. Series.
KJE6554.L64   1998
341.7'54—dc21   98–21463
                                                              CIP

British Library Cataloguing-in-Publication Data
A catalogue record for this book is available from the British Library.

The publisher offers special discounts on bulk orders of this book.
For information, please contact:

Manager of Special Sales
Butterworth–Heinemann
225 Wildwood Avenue
Woburn, MA 01801-2041
Tel: 781-904-2500
Fax: 781-904-2620

For information on all Butterworth–Heinemann publications available, contact our World Wide Web home page at: http://www.bh.com

10 9 8 7 6 5 4 3 2 1

Printed in the United States of America
Transferred to Digital Printing 2004

For
Tricia, Traci, Tom,
and
Carol

Contents

Preface

Welcome to *CE Marking Handbook: A Practical Approach to Global Safety Certification!* If you are a manufacturer or exporter of products bound for the European market you might have assumed that shipping products to Europe was a rather simple operation, but found yourself confused by the complex maze of European Conformity laws and regulations. If you are a consumer or consumer rights advocate you might have assumed that products on the European marketplace displaying the CE symbol guaranteed that the product had undergone stringent safety and electromagnetic compatibility (EMC) testing to have the right to bear this "coveted mark."

First of all the term *CE mark* is incorrect. It was amended in the directives to *CE marking,* which differentiates the suppliers self-declaration (CE) from a approval mark (VDE/TUV) issued by an accredited European Union (EU) certification body. Furthermore, it is not accurate to refer to CE as a self-certification as it is really only a self-declaration and no more. Therefore, it will always be referred to it as CE marking, as used in the directives themselves. It is a symbol of the supplier's self-declaration and not a certification, approval, or mark of any kind.

There is no such thing as a CE approval or CE certification! CE is *not* a *mark* or *approval,* it is a *marking* which is only a self-declaration under the suppliers' own responsibility.

CE Marking Handbook: A Practical Approach to Global Safety Certification strives to clear up the confusion surrounding CE marking. With CE marking now mandatory for most equipment (products, appliances, machines, etc.) bound for Europe, it is paramount that you, the manufacturer or exporter, gain a thorough understanding of the *New Approach* now in effect (after the transition period from the *Old Approach* has ended) to better ensure that your product is not waylaid by unexpected regulations and approvals. Informed consumers and consumer advocates also will be on the lookout for products. As of this writing, most of the legislation and certification procedures are now in place and the goal of "total harmonization" is, for all practical purposes, complete. Furthermore, numerous product safety, machinery, and EMC standards have been published to support the directives' essential requirements. The standardization process is evolutionary, with established procedures to cover all compliance aspects—even where a standard is lacking or not yet formulated. Knowing that no single book can cover all topics or products, the scope of *CE Marking Handbook: A Practical Approach to Global Safety Certification* is purposely limited to cover the subjects of most interest to equipment suppliers and consumers rights activists. Concerns, such as product safety requirements, conformity assessment procedures, and enforcement are covered. Electromagnetic compatibility and machine hazards are also discussed within the body of this book in the final chapters. *CE Marking Handbook: A Practical Approach to Global Safety Certification* does not portend to cover all aspects of the topic at hand, but endeavors to elucidate what is often confusing, provide practical guidelines, and dispel the myths surrounding CE marking.

Some people in industry feel that CE marking is just another trade barrier, but this is not the case since everyone, even Europeans themselves, must now follow the same rules. Further negating the notion of the CE Marking as being a trade barrier is the fact that in the past the member states of the EU had numerous barriers within Europe. Each country was setting its own laws and standards, and enforcement was not uniformly applied. The initial goal of the New Approach was total harmonization of regulations to remove Europe's internal barriers to trade, which was later expanded to a broader global approach. The objective was to ensure a more even and fair competition among member states and to improve competitiveness with products coming from outside Europe. The total harmonization concept has had a very positive effect in establishing one set of conformity laws, standards, and procedures to be adhered to universally. Furthermore, Europe's product standards and assessment procedures are now recognized internationally by most countries as the worldwide standards, making total harmonization a truly global approach.

To the product designer, manufacturer, exporter, and informed consumer, the CE marking raises many unanswered questions. Understanding the real meaning of the CE marking and what's required to affix it to products can take many hours, weeks, or months of study. Oftentimes, with so much conflicting information in abundance, those that took the initial steps to understand the CE marking process ended up more confused than when they started.

There is an abundance of myths surrounding the CE symbol and its value. The value of the CE marking for selling products has been overstated and the risks of nonconformity have not been well understood, and that accounts for much of the confusion. The CE marking is only the product manufacturer's symbol of self-declaration; it indicates a product's conformity to the minimum requirements of the applicable directives. Although the CE marking does permit a product's access to the EU, it is not an approval mark, certification, or quality mark; nor should it be a marketing tool. The CE marking is limited in scope. While enabling products to be placed on the European market it allows for the free movement of goods and permits the withdrawal of nonconforming products—but no more. There is no such thing as a CE approval or CE certification! In reality, the CE marking is not an approval mark or certification but is actually an *unqualified* mark, since it is only the manufacturer saying, "I did it—*trust me!*" Professionals and seasoned consumers familiar with products bearing only the CE marking sometimes refer to it as the "mark of nonconformity" because of their negative experiences with nonapproved products and components bearing this marking. The CE marking should not be confused with other approval marks or certifications issued by an EU-accredited certification body. As stated in the European Commission's *Guide to the Implementation of Community Harmonization Directives:* "Manufacturers are responsible for ensuring that the products they place on the market meet all relevant regulations. Where these regulations do not require mandatory certification, manufacturers often seek voluntary certification to assure themselves that their products do meet the requirements set by law."

Opinions and misinformation abound! All you have to do is ask a question and you will most likely get conflicting answers. To further complicate matters, the answer depends on who is asking the question. Asked by those involved in the legal technicalities of product design and exportation, the most asked question is "What do I *need* to 'ship' products to Europe?" The answer seems obvious—it's the CE marking! From the legal point of view, the CE symbol and proper documentation are all that's mandatory for access to the European Union in most cases, since it is *mandated by customs inspectors and enforcement authorities.*

Some manufacturers and marketing professionals overstate the importance of the CE marking by placing far too much emphasis on its importance and elevating its stature to one of covenant. By stating to the consumer that their products have the coveted CE approval, they expect the buyer to look no further and purchase their products without any independent official evidence of conformity.

For those involved in developing marketing strategies for the European market, the most asked question is: "What do I need to 'sell' products in Europe?" In this case product reliability, verified safety, and quality become paramount, since meeting these criteria are what is needed to gain *consumer acceptance.* The CE marking does not guarantee these factors. So *need* really depends on your perspective, that is whether you are trying to just *ship* the product or to *sell* the product. As stated by Mr. Bangemann, VP of the European Commission: "The CE marking is not a quality marking, although it is often wrongly perceived as such and then compared to other quality marks. Quality markings [approval marks], as opposed to the CE marking, are voluntary, address consumers or users, and tend to influence their appreciation toward the relevant product."

CE Marking Handbook: A Practical Approach to Global Safety Certification is a guide to understanding the why's and how to's of meeting European requirements for CE marking. The why's (or reasoning) are explained in the laws, known as directives; and the how to's (or product requirements) are explained in the technical rules, known as *harmonized standards.* The directives and standards must be considered together with a *focus on standards* for the product's design and assessment criteria.

The most important consideration must be on protecting the consumer (user, operator, buyer, etc.), which in turn limits risks to the product manufacturer and supplier. The product assessment procedures and design requirements are well established and defined within Europe's New Approach. However (in the few cases where there may be doubt), it is always prudent to err on the side of consumer protection. Only those seriously interested in limiting their risks by putting consumer protection first should read on.

CHAPTER 1

Europe's Approach to Total Harmonization

Every man has a right to his opinions.
But no man has a right to be wrong about his facts.

BERNARD BARUCH

The European Union and the CE Symbol

The European Union's (EU) goal of single-market access has finally arrived! For the most part it appears to be working! The EU's New Approach has reduced or in many cases eliminated the legal and technical differences between the member state countries that existed under the Old Approach.

The CE marking is a symbol of the manufacturer's self-declaration indicating that the product or machine conforms to all relevant European regulations. The requirements are contained in the various European directives (laws) and standards (tests and design). Directives, along with the appropriate European standards, cover all safety, electromagnetic compatibility (EMC), health, and environmental concerns. CE marking of machinery came into effect on January 1, 1995, and EMC conformity has been required since January 1, 1996. Electrical products must bear the CE marking as of January 1, 1997, for safety. The essential requirements (ERs) for product safety have been in effect since 1973.

This book is based on three primary directives, as amended by 93/68/EEC adding the CE marking:

- Low-Voltage Directive (LVD) 73/23/EEC (product input; 50-1,000VAC or 75-1,500VDC)
- Machinery Directive 89/392/EEC (includes mechanical and electrical safety)
- Electromagnetic Compatibility (EMC) Directive 89/336/EEC (LVD products and machinery)

After working with numerous companies in the field of European product safety and EMC compliance, I believe that confusion still surrounds the CE marking in countries outside of the European Union—especially in the United States. The value of the CE marking for selling products has been overstated and the associated risks of nonconformity have been, in general, not well understood. The actual purpose of the CE marking is to allow products to be placed on the market and ensure the "free movement of goods." With the exception of some high-risk products, most products can be self-assessed by the manufacturer to meet the ERs of the directives. The CE marking and a document called a declaration of conformity imply conformity with all relevant directives. Final acceptance of the product, however, is driven by the market (consumers).

In addition to the CE marking affixed to a product, customers may demand a European third-party (notified body) approval mark, certification, or test report connoting a higher level of quality to ensure safety/EMC compliance. To control nonconforming products the following clause from the *Official Journal of the European Communities (OJEC)* applies:

> A single CE marking should be used in order to facilitate controls on the Community market by inspectors and clarify the obligations of economic operators [manufacturer or supplier] . . . ; the aim of the CE marking is to symbolize the conformity of a product . . . and to indicate that the economic operator has undergone all the evaluation procedures laid down by Community law in respect of his product. (OJEC; 93/465/ EEC)

The main goals of the CE marking are to:

- Indicate a product's conformity to the "essential requirements" of the directives,
- Allow products be "placed on the market,"
- Ensure the "free movement of goods," and
- Permit the "withdrawal of non-conforming products" by customs and enforcement authorities.

It is well advised that manufacturers and suppliers be informed not only of the CE marking's advantages, but also of its limitations. The CE marking:

- Is *not* a mark or certification or approval issued by a third party,
- Is *not* for sales or marketing or promotion,
- Is *not* a quality mark, and
- Is *not* for components (although there are some exceptions, the vast majority of components do not need CE marking; see Design Guide— Components in Chapter 6).

This book does not address all regulatory aspects for medical devices, telecom products, RF transmitters, or other regulated products. To reduce the number of equivalent or similar terms, the following terms are considered interchangeable:

product/machine/equipment; manufacturer/supplier; notified/competent body. In general, the terms *third party* and *notified/competent body* are considered equivalent and carry the same meaning in Europe. Thus, European accredited bodies (third parties) are sanctioned to perform various assessment functions, such as testing and/or certification.

Definitions and Considerations

Most all products must be assessed for conformity by either the manufacturer or an accredited body and bear the CE marking before entry into the EU. The CE symbol indicates that the product conforms to the relevant directives and that the manufacturer has performed all the necessary assessment procedures. Neither the quantity of products or machines nor the nature of the transfer relieves the manufacturer of performing the conformity assessment and affixing the CE marking. The marking applies to all equipment put into service in a public or private capacity for professional or nonprofessional use or paid for or free of charge. The key points discussed in this guide will be:

- Review of the EC directives—Gain understanding of the legal and procedural issues, such as the essential requirements, procedures, enforcement, and rules for CE marking.
- Discussion of compliance with European *harmonized standards*—This is the most important element to meeting the ERs of the directives. Products manufactured in conformity with harmonized standards are "presumed to conform" to the ERs of the directives.
- What's needed in conjunction with affixing the CE Marking—With the product supplier's CE marking mandatory for most products, the supplier's declaration of conformity and technical file must also be readily available on demand by enforcement authorities.
- Importance of voluntary certifications, test reports, and approval marks—Although it is not mandatory for most products and machines, voluntary certifications, test reports, and approval marks issued by EU certification bodies are the best means of showing a *defense of due diligence* should a product become suspect by enforcement authorities, competitors, or customers. Due diligence means taking all reasonable steps to ensure conformity. Regardless of the presence of the CE marking to achieve a higher level of quality and ensure safety/EMC compliance, customers can also demand an approval mark and/or test report from an EU third party.

To understand European Conformity, we must first be familiar with a few important terms. The term *equipment* used in the standards can mean almost any product, appliance, or machine and, therefore, these terms are interchangeable. The term *product* is preferred by this author and will have the same meaning as equipment (e.g., product, appliance, or machine) and may include components in the general sense. *Components* are devices that are built into equipment, cannot function alone, and rely on the end product for their safety. Review the following definitions, as these will be the basis of our discussion.

Official Journal of the European Communities (OJEC). This is Europe's register of legislation, directives, information and notices, lists of standards, and notified bodies. Perhaps the most singularly important document for European Conformity, the *Official Journal* as it is often called, is where the directives and standards are published, making them valid for use by product manufacturers and accepted by national authorities. Amendments to the directives and standards are included in the OJEC as well as listings of the accredited testing and certification bodies. The *OJEC* is published and revised on a regular basis.

Directives are the European laws published in the *Official Journal* that give us the essential health and safety requirements (EHSRs) that shall be followed by product suppliers. These are commonly called the Essential Requirements (ERs). The directives deal with legal and procedural issues, such as assessment procedures, certification, implementation, enforcement, technical files, declarations, CE marking, and other basic concepts. Examples of primary directives are the Low-Voltage, Machinery, and EMC directives. The General Product Safety and Product Liability Directives are basic directives dealing with enforcement and civil prosecution that are applied to all products. The directives also mandate the publication and mutual recognition of harmonized standards.

Harmonized standards provide the specific technical rules for product design and conformity testing. Standards become harmonized and valid for use once they are published in the *Official Journal.* Harmonized standards, known as European Norms (ENs), give the detailed engineering requirements (safety/EMC) for products such as construction materials, wiring, components, labeling, warnings, documentation, and testing and pass/fail criteria. Conformity to the harmonized standards is recognized as the minimum acceptable criteria. However, other requirements may also apply. Harmonized standards (ENs) are the cornerstone of the Europe's New Approach and without them the system would not function.

Presumption of conformity is a phrase associated with products that comply with all aspects of the relevant European harmonized standards. National enforcement authorities are obliged to recognize that equipment manufactured in conformity with the harmonized standards, published in the *Official Journal* and transposed into national standards, are presumed to conform to the essential requirements of the directives. When a product is not in compliance with the appropriate European standards, the equipment will not benefit from a presumption of conformity conferred by the use of such standards.

Manufacturer is the entity responsible for designing and manufacturing a product that is covered by a directive or directives with a view to placing it on the market. The manufacturer shall follow the appropriate assessment procedures and is responsible for all aspects of the product's conformity. This includes, but is not limited to, the product's design, testing, reports, documentation, declaration, technical file, and CE marking. The manufacturer is ultimately liable for the product's conformity and the accuracy of the technical file. Manufacturers may, if they so choose, use an authorized representative for limited delegated responsibilities.

An *authorized representative* is the person appointed by a manufacturer to act on its behalf within the EU. There is no mandatory requirement to have a authorized representative for most products. The representative shall be established in the EU and is the person who signs the declaration, holds the technical file, and carries out certain tasks required by the directives, as agreed upon by the manufacturer.

Evaluation of conformity is an assessment and is the systematic review to which a product, or system, fulfills specific requirements (i.e., harmonized standards). A mandatory assessment according to all the requirements must be undertaken and documented in the manufacturer's technical file before anyone can affix a CE marking and place a product on the market, make it available, or put it into service.

A *technical file* is required by the primary directives to document the conformity assessment and the product's design. The technical file shall be compiled by the manufacturer or authorized representative and contain design documentation, manufacturing procedures, test reports, and operation information to show conformity as required by the directives.

A *technical construction file* is a special file for regulated products, high-risk machines (Annex IV machinery), and occasionally utilized for EMC (large equipment, etc.), where the use of a notified or competent body is mandatory.

Test reports are technical records on the conformity assessment of a product according to specific standards. Test reports are concise accounts, including clause-by-clause details on the results of the product safety/EMC assessment, standards rationale, test data, construction, and critical components. Test reports are an essential tool for conformity assessment. Test reports contain the technical results of the conformity assessment and need not contain confidential design information. Test reports may be requested by customers, testing/certification bodies, or enforcement authorities for review and verification purposes.

A *declaration of conformity* is drawn up by the manufacturer or authorized representative and used to declare that the product has undergone all the necessary assessments and that the product satisfies the essential requirements of the applicable directives. This self-declaration lists the standards and directives applied, the responsible person and other necessary information as required by the directives.

CE marking is the manufacturer's or supplier's self-declaration symbol to indicate that the product has undergone all the necessary evaluation procedures and is in conformity with the minimum requirements of the relevant directives. Products that conform to all the applicable directives and standards may bear the CE marking. The CE symbol is not a registered trademark and is, in principle, under the manufacturer's control. Products bearing the CE marking may be placed on the market.

National authorities enforce European and member state laws at the national level. Local authorities have the responsibility for market control through product checks based on incidents, complaints, or random audits. The authorities take appropriate measures to control nonconforming products, including withdrawal, fines, and possible imprisonment. In some states, such as Germany, the authorities may act without having to go to court. The national authorities are obliged to inform the Commission and/or the other member states who then also must take the appropriate action.

Accreditation bodies are independent national accreditors who evaluate testing and certification parties according to recognized norms. After a third party is found competent in a member state, such as in Germany, the accreditation body then notifies the European Commission who notifies the other member states by listing them in the *Official Journal*. These third-party testing and/or certification bodies are then considered EU-accredited third parties or notified bodies.

Notified bodies are independent testing laboratories and/or certification bodies recognized in the European Union to perform tests and issue reports and certificates on conformity. These bodies are generally referred to as notified bodies. Depending on the directive an accredited body may be referred to as a notified body, competent body, certification body, third party, or other (see Notified Bodies and Third Parties in Chapter 4). Test reports and certificates issued by accredited bodies attest to a product's or system's conformity to the relevant standards. These reports and certificates are the basis for mutual recognition of test results and build consumer confidence in a product's conformity.

Certification and approvals are statements by an impartial accredited body that a product or service fulfills specific requirements, such as directives and standards. Certification may be mandatory, as is the case for some regulated products (Annex IV machines, telecom, medical devices, etc.), or voluntary as is the case for most product and machine categories. Manufacturers often seek voluntary certification to assure themselves, customers, and authorities that their products meet the requirements set by law. Certification is commonplace in Europe and allows the use of a distinctive approval mark affixed to the product that is backed by a certificate and test report. The "approval mark," is a recognized quality mark attesting to a products conformity to the relevant requirements, such as for EMC and safety.

💡 Directives tell us *why* we must comply (consumer safety/EMC) and *what* may happen if we ignore the laws (withdraw products). But it's the European *standards* that show us *how* to comply (design and assessment).

History: The German Model

The history of occupational safety in Europe dates back to the turn of the century with safety laws enacted in some countries, such as Germany. The Accident Insurance Act (UVV) of July 6, 1884 was the first of its kind in the world and led to fundamental change. Previously, compensation for an industrial accident or an occupational disease had to be sought directly from the company itself—often a pointless undertaking since it had to be proved that the employer was at fault. So a social security system was created from health insurance schemes, pension funds, and Trade Cooperative Associations (BG) with the intention to give manual and white-collar workers material support in the event of sickness, old age, or accident at work. A special feature of the German accident insurance system is that compensation for an industrial accident or occupational disease can be claimed from the BG, who assumes liability in place of the individual companies. Some of the Accident Prevention Regulations (VBG), and safety codes and guidelines (ZH), issued by the BG, apply to all branches of industry. Others address particular machines, facilities, or sectors of industry such as office workplaces. The Accident Prevention Regulations also serve the implementation of national laws having the status as statutory ordinances. In addition to accident insurance and safety regulations the BG is a enforcement body in charge of monitoring company compliance with accident prevention regulations and workplace protection standards. The BG inspectors advise manufacturers on the safety and ergonomic design of tools and machines and they are involved in safety checks on products and equipment.

As is commonly known, Germany has strict safety/EMC laws and enforcement. Through the German Ministry of Labor, safety/EMC laws are passed by voluntary consensus with the support of the general populace, consumer protection groups, and the BG. These laws pave the way for the legality and recognition of standards. Historically, the Deutsches Institute for Normen (DIN) has worked to establish standards to satisfy society's global needs. Besides satisfying the needs of the enforcement authorities, of paramount importance are the needs and protection of the user. The DIN informs the state authorities of any standardization work and completed standards. In the past, the standards were typically released as DIN or DIN-VDE standards, but now DIN checks how the newer DIN-EN standards and the EU directives translate into German law and use by the authorities. In fact, the German safety and RFI (EMC) laws, as well as the standards, have served as models for the New Approach and are now adopted at the European Union level. The German RFI law (HfrG) of 1949 required conformity to various RFI legislation and standards for compliance. European and international working groups modified these German RFI laws and released them as European EMC laws and standards (CISPR). The same process occurred for safety, with the German Equipment Safety Law (GSG) being the model for Europe's Low-Voltage and Machinery Directives.

As an example of standards history and progression, the German ITE safety standard DIN VDE 0805 became EN 60950 (IEC 950), machine safety standard DIN VDE 0113-1 became EN 60204-1 (IEC 204-1), machine EMC standard DIN VDE 0875 became EN 55011 (CISPR 11), and appliance EMC standard DIN VDE 0875-1 became EN 55014 (CISPR 14). Except for EMC immunity, the majority of European product safety, machinery, and EMC standards are traceable back to their German standards origin. Thousands of German standards (DIN, VDI, VDE, VBG, ZH, etc.) form the basis for the newer harmonized standards (EN, IEC, etc.).

The legality of standards, as the recognized technical rules for equipment safety was first established at the German national level with the GSG, which states that "equipment shall be constructed in accordance with the generally accepted rules of technology [mostly standards]." This national requirement of utilizing standards and specifications carries over into the New Approach as the recognized method for achieving the *presumption of conformity*. Conformity with the EU harmonized standards (EN) and their national equivalent (i.e., DIN-EN), therefore, constitutes the basis for a product's presumed technical legitimacy and are in practice the only verifiable path to ensure conformity, limit the producer's risks and liability, and meet the authority's and consumer's expectations.

Some people believe that the reason for the popularity of German safety approvals such as the GS Mark, is that approvals were mandatory in Germany. After all, over 100,000 different products bear the GS Mark (GS = Safety Tested). Approvals and marks have, however, never been mandatory by German law for the vast majority of products and machines. Their popularity is driven by expectations in the marketplace from the product users, consumer groups, insurers, and employers. These high expectations continue today in Germany and some other countries. (See chapters on Notified Bodies and Certification [Chapter 4] and the Quality and Safety Mindset [Chapter 5].)

The New Approach: 1985

Prior to 1985, trade, quality, safety, EMC, and other areas differed among many EC (now EU) states. Each country established its own laws and standards and even when the EC attempted to harmonize standards national deviations existed that limited the reform efforts. These internal restrictions not only made it difficult for European Community States to trade with other EC countries, but also the restrictions were beginning to leave Europe behind in the world market. Europe needed to reduce the cost of its products and services to compete in the global market and maintain acceptable employment levels. Harmonization started in 1957 with the Treaty of Rome and the founding of the European Economic Community (EEC). Even so, competition from overseas was outpacing the reforms, and standardization in Europe was moving too slowly. To achieve a unrestricted free market within Europe, drastic action had to be taken and fast. One action was the removal of technical barriers to trade, which then became a major factor in the promotion of industrial quality and of the competitiveness of European firms—both on the internal market and beyond. So, a principle of European directives (laws) and standards (technical rules), called the *New Approach,* was introduced in 1985 and has gained momentum ever since. Figure 1-1 shows a history of EC directives.

Directives issued before 1985 are considered Old Approach directives except when amended after this date, as is the case for the Low-Voltage Directive (LVD) 73/23/EEC, amended by 93/68/EEC adding the CE marking. The LVD is valid under the New Approach.

1872	TUV founded in Germany for industrial safety.
1884	Accident Insurance Act enacted in Germany for Insurance and Trade Cooperative Associations (BGs).
1893	VDE founded in Germany for safety of electrical products.
1906	IEC established for safety criteria in electrical industry and international trade.
1949	German RFI Law (HfrG) implemented for electrical products (limits RF emissions).
1957	Treaty of Rome. Initiates the "single market" concept for free trade.
1968	German Equipment Safety Law (GSG). All products shall be safe per relevant standards.
1973	CENELEC formed to coordinate standards activities.
1973	Low-Voltage Directive 73/23/EEC issued for electrical product safety.
1985	New Approach initiated for harmonized directives and standards.
1987	Single European Act. Area without borders for free movement of goods, personnel, etc.
1989	Global Approach supplement to New Approach for conformity assessment and CE Marking.
1989	EMC Directive 89/336/EEC issued for electrical products, appliances, machinery, etc. (immunity and emissions).
1989	Machinery Directive 89/392/EEC issued for machine safety.
1992	Treaty of Maastricht for one policy of defense, foreign affairs, citizenship, and currency.
1993	CE Marking Amendment 93/68/EEC issued for CE marking, declaration, and technical file.
1993	Treaty on European Union completes the process of the internal market policies.

FIGURE 1-1   A History of European Directives

Under the New Approach the entire process was placed on a fast-track, forcing the release of many new directives and standards, as well as amendments to existing directives and standards to bring them in line with the New Approach. This approach mandates a "total harmonization" concept through the implementation of European laws known as *directives*. The purpose of the New Approach is to establish one set of regulations (directives, standards) for all to follow to ensure the "free movement" of goods, persons, services, and capital. The rules are therefore to be consistent and applied uniformly throughout the Community. When the regulations are applied properly, manufacturers may affix the CE marking to products, which allows products to be "placed on the market."

Another new strategy called the Global Approach is a supplement to the New Approach, establishes uniform conformity assessment procedures and rules for the CE conformity marking. One of the goals of the Global Approach is to harmonize conformity assessment to create the conditions essential for engendering confidence and, hence, the establishment of mutual recognition. Consequently, prior to affixing the CE marking to products, manufacturers must either conduct conformity assessment on their products by themselves if they are competent and properly equipped or subject their products to conformity assessment by specialist third parties. These specialist third parties consist of European testing laboratories and certification bodies commonly known as notified/competent bodies (safety/EMC).

1985–The New Approach
- Harmonized directives (essential requirements)
- Harmonized standards (technical rules)

1989–The Global Approach
- Conformity assessment procedures (the modules)
- CE marking rules (manufacturer's self-declaration)

Products must conform with the Essential Requirements (ERs) of all relevant directives before being placed on the market. ERs are at the center of European law and standards and are retained for the presumption of conformity by virtue of their publication in the directives as harmonized standards. The ERs and harmonized standards include the requirements to protect consumers, health, merchandise, and the environment. The transposition of the European directives and harmonized standards into national laws and standards makes them legally binding and obligatory in the member states. Products that conform to the European harmonized standards are presumed to comply with the essential requirements of the directives.

Compliance with the Essential Requirements via European standards is generally considered the *minimum* acceptable criteria.

CE Marking Directive 93/68/EEC states that "the choice of conformity assessment procedures must not lead to a lowering of safety standards of electrical equipment/products, which have already been established throughout the Community." Alternatives to European standards are possible in the case where no European harmonized standard exists. Conformity to International standards (i.e., IEC) or European national standards (i.e., DIN, BS, NF) is sometimes an acceptable alternative, but in this case the manufacturer must be able to demonstrate conformity with the ERs of the directives and at the same time not fall below the minimum acceptable criteria. The General Product Safety Directive 92/59/EEC allows the use of EU national standards when no European harmonized standards exist. This alternative may be needed as a transitional measure until a European standard is published (see General Product Safety in Chapter 2). This approach is not recommended when European harmonized standards exist, especially if a product becomes suspect, since one can expect the national enforcement authority will use the EU harmonized standard, or their national equivalent, to evaluate a product's lack of conformity. Going against the directives, authorities, and consumer expectations (see Product Liability in Chapter 2) may place the product's manufacturer in a hopeless situation if the product's conformity is questioned.

When European standards (ENs) exist, always use them to ensure conformity with the European directives. In some cases European alternatives (e.g., EU national standards) may be considered but must be justified. Use of other so-called non-European alternatives, such as U.S. standards and others, may be indefensible in court and could actually place the product and manufacturer in a position of nonconformity.

Standards versus Directives for Conformity

Europe's directives (laws) describe the *Essential Health and Safety Requirements* that suppliers must meet before equipment is placed on the market. The directives tell us *what* must be done from a procedural and legal point of view, such as the CE marking, the declaration, the technical file, enforcement, or design concepts. Standards, on the other hand give detailed safety requirements that tell us exactly *how* safety and EMC may be accomplished in the field of engineering and design, such as guarding, warnings, components, electrical, testing, or pass/fail criteria. Various machine safety concepts and documentation requirements are also discussed in the Machinery Directive, but they are presented for the most part in general terms. The specific technical rules, tests, and pass/fail limits for safety and EMC are detailed in the relevant harmonized standards

There are several ways for product and machine manufacturers to show compliance with the European directives. However, only one way *presumes conformity*— the use of harmonized European standards. This is why I strongly recommend the application of harmonized standards, also known as European Norms (ENs). Standards play an important role in European Union safety/EMC compliance for products and machines. Standards become valid in Europe after they are published in the *Official Journal* and are transposed into member states' national standards. The ENs are listed or appended to the directives themselves.

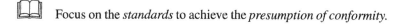 Focus on the *standards* to achieve the *presumption of conformity.*

Some directives, such as the *Global Approach to Certification and Testing,* state that a manufacturer's equipment must meet the ERs of the directive and that the use of harmonized standards is "voluntary." The voluntary clause dates back to the early white papers and basic directives when there were as yet few published standards for machinery. The voluntary concept was instituted to allow the use of other standards, such as IEC or national standards, until the ENs were published in the *OJEC.* This has now changed with numerous EN and IEC standards available.

The directive on quality and conformity assessment says "the economic operator is not obliged to follow the European standards referred to in the directives" (*OJEC*; 89/C 267/12) and

> that there could be more than one means of proof of conformity to a directive. It provided for presumption of conformity to a directive on the basis of a European standard, or, during a transition period, of national standards which have been submitted and recognized as equivalent under a Community control procedure. When the manufacturer complies with these standards, the directives are there to allow him to make use of simplified certification mechanisms. . . . When the product does not conform to a standard, however, either because the standards do not exist or because the manufacturer [in the case of innovation] prefers to apply other manufacturing criteria of his own choice, the assessment of conformity to the essential requirements must involve a third party [EU body], either by certification or by third party testing. (*OJEC*; 89/C 267/20)

Even though other means may be possible, the harmonized standards route is the simplest and lowest risk alternative. In practice the manufacturer should consider the use of harmonized standards advantageous to ensure their equipment's safety compliance and to establish a presumption of conformity. The directives state:

> National authorities are obliged to recognize that products manufactured in conformity with the harmonized standards [EN] published in the *Official Journal of the European Communities (OJEC)*, and transposed into national standards, are "presumed to conform" to the essential requirements of the directives.

When considering standards for a product safety assessment the appropriate standards shall be applied. They are, listed in order of preference:

1. A European harmonized standard (EN), published in the *Official Journal*;
2. An IEC standard that has been published in the relevant directive(s) when a "EN" does not exist;
3. When a harmonized EN or IEC does not exist, a EU national standard; and
4. Codes of practice recognized within the Community where no EN, IEC, or national standard exists.

The question of using U.S. standards for EU compliance always arises. Because there are numerous technical differences between the two countries' standards and the U.S. standards are not listed in the *OJEC,* these standards may not be considered for European conformity. Also, there are major differences in philosophy between the European Union and United States. The U.S. focus is primarily on "flammability and testing," whereas the EU focus is on "shock hazards and construction." Although both the U.S. and EU safety experts consider the same safety aspects and tests, it is with a different emphasis. This often means the difference between passing and failing. U.S. standards are not equivalent or comparable to the EU harmonized standards. The EU standards and interpretations are generally considered to be more strict (see U.S. and EU Differences in Chapter 5).

Without mutually agreed-upon standards, it would be difficult to establish conformity in quantifiable engineering terms to determine whether a product or machine meets any minimum safety levels or technical requirements. When harmonized standards are not used, a "presumption of conformity" does *not* exist. This places the manufacturer in a potentially indefensible position should their equipment become suspect by customers, competitors, or enforcement authorities. Some manufacturers, looking for loopholes, refuse to apply the harmonized standards or only apply them in part (pick-and-choose), saying they have shipped numerous products to Europe and they have not injured anyone. Be wary of this rationale as it does not establish conformity, but instead should alert the buyer to potentially nonconforming products. A product's "verifiable conformity to the harmonized standards" goes a long way to satisfy your customer's compliance concerns and is a good defense if your equipment's conformity is questioned by authorities.

CHAPTER 2

Directives and Procedures

The only thing more expensive than education is ignorance.

BENJAMIN FRANKLIN

Directives: The European Laws

European directives are laws that manufacturers must meet before affixing the CE marking to products. The directives are identified with the year and identifier number such as "85/375/EEC" for the Product Liability Directive, which was published in the *Official Journal* on August 7, 1985. Most directives are adopted at a later date, within a specified time limit. For example, article 19 of the Product Liability Directive 85/375/EEC states that this directive must be brought into force by the member states within three years.

The directives typically apply to:

- Member states
- Manufacturers
- Certification and standards bodies
- Importers, dealers, and users

As of this writing there are over seventeen generic and product-specific directives, as well as other important basic directives. Figure 2-1 shows the directives for the New Approach to date. Here is an overview of the three types of directives:

Basic directives (type A). The basic directives apply to all manufacturers of products and address trade, enforcement, liability, and other issues. It is important to understand the implication of these directives, especially concerning enforcement against product manufacturers. Examples of the basic directives are CE Marking, Conformity Assessment, General Product Safety, and Product Liability. Products, components, and materials not covered by Type B or C directives must still be safe according to the General Product Safety Directive (e.g., comply with standards).

Directive:	Ref No:	CE in force
Low-voltage directive (LVD)	73/23/EEC	1/1/1997
Electromagnetic compatibility (EMC)	89/336/EEC	12/31/1995
Machines	89/392/EEC	12/31/1994
Simple pressure-vessels	87/404/EEC	7/1/1992
Pressure equipment	97/23/EC	5/29/2002
Equipment for use in explosive atmospheres	94/9/EC	6/30/2003
Active implantable medical devices (AIMD)	90/385/EEC	12/31/1994
Medical devices—general (MDD)	93/42/EEC	6/15/1998
Medical devices—*in vitro* diagnostics (IVD)	COM(95)130	7/1/2002
Telecommunications terminal equipment and satellite earth station equipment	98/13/EEC	11/6/1992 or 5/1/1995
Non-automatic weighing machines	90/384/EEC	1/1/1993
Gas appliances	90/396/EEC	12/31/1995
Household appliances (energy efficiency)	96/57/EC	3/9/1999
Toys	88/378/EEC	1/1/1990
Recreation craft (small boats)	94/25/EC	6/16/1998
Boilers	92/42/EEC	1/1/1988
Construction products	89/106/EEC	6/27/1991
Personal protective equipment	89/686/EEC	7/1/1995
Passenger lifts	95/16/EC	7/1999
Explosives	93/15/EEC	1/1/2003
Directive reference numbers ending with EC were issued after 1993. Directive 98/13/EC replaced 91/263/EEC and 93/97/EEC.		

FIGURE 2-1 List of EC Directives for CE Marking

Generic directives (type B). Generic directives address a specific range or group of products, such as products operating between certain voltage limits under the Low-Voltage Directive (LVD), or products that may generate RF emissions (EMC). These directives cover the "unregulated" sector of products. Typical unregulated products are information technology equipment (ITE) or household appliances. Most prominent among the generic directives are the Low-Voltage and EMC directives. Note: In the case of EMC, where no standards exist or are not applied in full, involvement of a EU competent body may be required prior to CE marking.

Product-specific directives (type C). Product-specific directives apply to "regulated" products, such as telecom and medical, as well as other products where extreme hazards exist, such as machinery listed in Annex IV of 89/392EEC. Involvement of a EU body may be mandatory, but not in all cases. If the class or type of product is regulated by the directive, then a European body must assess it and issue a *Type-Exam Certificate* prior to CE marking. Product-specific directives cover machinery, pressure vessels, medical products, telecom devices, toys, and others. It is necessary to refer to the relevant directive to see whether the involvement of a EU-notified body is mandatory for the product or machine in question. Type C directives take precedence over types A and B directives and refer to them as needed.

I focus on three "primary" directives (types B and C) that relate to electrical products and machinery, as well as some of the basic directives (type A). Most electrical products and machinery fall under two primary directives and several basic directives. The three primary directives are:

The Low-Voltage Directive (LVD), 73/23/EEC. The LVD is an electrical safety directive and applies to products and equipment for which the hazards are primarily of an electrical nature and that operate between 50 and 1,000 Vac or 75 and 1,500 Vdc. The LVD's "essential requirements" have been in effect since 1973 and mandate conformity for all safety aspects of electrical products, including those of mechanical origin.

The Electromagnetic Compatibility (EMC) Directive, 89/336/EEC. The EMC Directive controls the emissions and immunity characteristics of electrical products and machines (called apparatus). Electromagnetic compatibility is the ability of an apparatus to operate satisfactorily in its electromagnetic environment. EMC is based on the two principles of limiting the electromagnetic energy (emissions) and affording adequate protection (immunity) against such energy occurring in the environment.

The Machinery Directive, 89/392/EEC. In brief, a machine is defined as an assembly of linked parts of at least one that moves and joined together for a specific application, for the processing, treatment, moving, or packaging of material. The directives' "essential requirements" cover all safety aspects of machinery, including mechanical, electrical, and others. The Low-Voltage Directive is not needed for machinery, since compliance with electrical safety is also required by the Machinery Directive (refer to standards; EN 60204-1 and others).

📖 *Helpful hint!* If you have trouble determining whether the Low-Voltage or Machinery Directive applies to your equipment, identify the appropriate product or machine safety standard first. It is then easy to know which safety directive applies since product standards are usually listed under only one of the directives.

Some of the basic directives (type A) that apply to most product and machine suppliers are:

- General Product Safety Directive—92/59/EEC
- Product Liability Directive—85/374/EEC
- Conformity Assessment Procedures and CE Marking Rules—93/465/EEC
- CE Marking Amendment—93/68/EEC

General Product Safety: Protection and Safeguards

The General Product Safety Directive 92/59/EEC requires product manufacturers and suppliers to "place only 'safe products' on the market" and to ensure a "high level of protection of safety and health of persons." The directive also gives member states authority to establish national "enforcement authorities." Authorities have the power to monitor product compliance (or lack thereof), take appropriate measures, and impose suitable penalties. When necessary, the member states shall immediately and efficiently organize the removal of dangerous products from the market and inform the Commission of the restriction or withdrawal. If justified, the Commission then notifies the other member states who also take appropriate actions. When necessary, the Commission informs consumers of the risks posed by the dangerous product(s). The severity of the action depends on the situation and nonconformity of the product. Directive 92/59/EEC replaces 89/45/EEC and incorporates its system of "rapid exchange of information" on dangers arising from products at national and European levels. See Figure 2-2 for safety terms.

In addition to supplying safe products, producers must also provide consumers with the relevant information to assess the inherent risks of the product where risks are not immediately obvious without adequate warnings. Such warnings, however, do not exempt any producer from compliance with the other provisions of this directive.

Based on the information in their possession, distributors shall act with due care to ensure safety compliance, in particular by not supplying products that they know or should have presumed do not comply with the safety requirements. The producers and distributors themselves must be informed of the risks products pose and adopt measures to enable them to take appropriate action including, if necessary, withdrawing the product in question from the market. Products must be marked and traceable for monitoring purposes to investigate complaints, keep distributors informed, and so on.

When a product has been proved dangerous producers must be able to locate the product and withdraw it from the market or, if necessary, organize its destruction.

a. *Product* shall mean any product intended for consumers or likely to be used by consumers, supplied whether for consideration or not in the course of a commercial activity and whether new, used, or reconditioned.

b. *Safe product* shall mean any product which, under normal or reasonably foreseeable conditions of use does not present any risk or only minimal risks compatible with the products use, considered as acceptable and consistent with a high level of protection for the safety and health of persons, taking into account:
   —the characteristics of the product, including its composition, packaging, instructions for assembly, and maintenance,
   —the effect on other products, where it is reasonably foreseeable that it will be used with other products,
   —the presentation of the product, the labeling, any instructions for its use and disposal, and any other indication or information provided by the producer,
   —the categories of consumers at serious risk when using the product, in particular children.

c. *Dangerous product* shall mean any product which does not meet the definition of "safe product."

d. *Producer* shall mean:
   —the manufacturer of the product within the Community and any other person presenting himself as the manufacturer by affixing to the product his name, trademark or other mark, or the person who reconditions the product,
   —the manufacturer's representative, when the manufacturer is not established in the Community or, if there is no representative within the Community, the importer of the product,
   —other professionals in the supply chain, insofar as their activities may affect the safety properties of a product placed on the market.

e. *Distributor* shall mean any professional in the supply chain whose activity does not affect the safety properties of the product.

FIGURE 2-2   General Product Safety Terms

When there are no specific Community provisions for safety of a product (i.e., EU laws and standards), a product may be deemed safe when it conforms to the specific rules of national law of the member state (i.e., national laws and standards). In the absence of Community and national rules for safety, then conformity may be based on "national standards" that are equivalent to a European standard or, where they exist, to Community "technical specifications," or, failing these, "member state standards" or "codes of good practice" in the sector concerned or to the "state of the art and technology" to the safety that consumers may reasonably expect [see Standards in Chapter 3].

Even when a product is in conformity with the relevant standards, authorities are not barred from taking appropriate measures to impose restrictions to marketing of a product, or require its withdrawal where there is evidence that despite such conformity, it is dangerous to consumers (see Standards and Essential Requirements in Chapter 3).

The member state authorities have the necessary powers to (Art. 6):

1. Organize appropriate checks on the safety of properties of products, even after being placed on the market as being safe;
2. Require all necessary information from the parties concerned;
3. Take samples of a product or a product line and subject them to safety checks;
4. Subject product marketing to prior conditions designed to ensure product safety and requiring that suitable warnings be affixed regarding the risks that the product may present;
5. Make arrangements to ensure that persons who might be exposed to a risk from a product are informed by publication of special warnings;
6. Temporarily prohibit while carrying out the various checks anyone from supplying or exhibiting a product whenever there are precise and consistent indications that it is dangerous;
7. Prohibit placing on the market a product or batch that has been proved dangerous; and
8. Organize the effective and immediate withdrawal of a dangerous product or batch already on the market and, if necessary, its destruction.

Where a member state takes measures to restrict the placing of a product on the market or require its withdrawal, the member state shall inform the European Commission of the measures, specifying its reasons for adopting them. The Commission shall then enter into consultations with the parties concerned. If the Commission concludes that the measure is justified, it shall immediately inform the member state that initiated the action and the other member states. Where the Commission concludes that the measure is not justified, it shall immediately inform the member state that initiated the action.

Member states may adopt "emergency measures" when a product poses a serious and immediate risk. In this case the Commission, with the assistance of a Safety Committee, and the member state cooperate to expedite the actions according to the urgency of the matter. The Committee's opinion shall be delivered to the Commission in a timely matter not to exceed one month. The Commission shall then adopt the measure, and the member states shall take all necessary measures to implement the decisions within ten days. Measures requiring the withdrawal of a product from the market shall consider the need to encourage distributors, users, and consumers to contribute to the implementation of such measures.

The member states and the Commission shall not disclose information obtained for the purposes of this directive, except for information relating to the safety properties of a given product that must be made public to protect the health and safety of persons. Parties concerned with a decision involving a product restriction may challenge any decision before the competent courts.

Any decision taken by virtue of this directive shall be without prejudice to Product Liability directive 85/374/EEC and without prejudice to the assessment of the liability of the party concerned, in the light of the national criminal law applying in the case in question (see Product Liability).

In addition to the enforcement provisions contained in 92/59/EEC on General Product Safety, the primary directives (Low-Voltage, EMC, and Machinery directives) clearly state that a member state may restrict a product's marketing or take other appropriate actions for any of the following reasons:

- Failure to satisfy or nonconformance with standards (European standards), or
- Faulty application of standards (misinterpretation or standard used has shortcomings), or
- Failure to comply with good engineering practice, or
- Product liable to endanger the safety of persons, or
- Product does not comply with the protection requirements (EMC), or
- CE marking has been affixed unduly.

When a member state ascertains that a product does not comply with the requirements, it may invoke the "safeguard" clause included in the directives and require the manufacturer to end the infringement under conditions imposed by the state. The manufacturer/agent shall be obliged to:

- Restrict or prohibit the product(s) sale, and/or
- Make the product(s) comply, and/or
- Withdraw the product(s) from the market, and/or
- Destroy the product(s).

Measures taken by the authorities shall be addressed, as appropriate, to:

- Producer;
- Distributors and/or party responsible for the first stage of distribution, within their limits of activities; and
- Any other person, where necessary, with regard to cooperation in action taken to avoid risks.

The "safeguard" clause included in the primary directives allows the total ban of nonconforming products and possible fines and/or imprisonment of the responsible person.

Finally, CE Marking Amendment 93/68/EEC reiterates the theme;

where a Member State authority establishes that a CE marking has been affixed unduly, the manufacturer/agent shall be obliged to make the product comply as regards to the provisions concerning the CE marking and to end the infringement under conditions imposed by the Member State. Where non-compliance continues, the Member State must take all appropriate measures to restrict or prohibit the placing on the market of the product in question or to ensure its withdrawal from the market.

Note: This section focuses on key elements of the directive and is only a part. Refer to the directives for the complete text. The actual wording of the directives is binding.

Product Liability: Victims' Rights

Prior to the New Approach directives, should a manufacturer find itself in court, the manufacturer need only show that reasonable measures were taken to ensure a safe product. In 1985 the Product Liability Directive was issued, changing the old focus of "negligence" to that of "strict liability," on the manufacturer's or supplier's part, thereby shifting the onus better to protect the consumer and placing a greater burden on the product supplier. Strict liability means that in the case of a claim the producers of a defective product can no longer exonerate themselves by proving "no fault." Even a defense of "due diligence" may no longer protect the producer from liability. The due diligence defense is still to have some importance, although it may be more difficult to prove. Since the negligence (no fault) defense no longer exists, the notions of "defectiveness, state of the art, and consumer expectations" are of new importance. "Quality control, documentation, and warnings" also gain increased significance. With respect to safety/EMC conformity, producers must document their quality control procedures and have a duty-to-warn consumers of hazards.

Since the Product Liability Directive was implemented, a trend toward "American-style liability" has resulted and civil lawsuits are on the rise in Europe.

Strict liability applies to "producers or suppliers" of any end product, raw material, or component, as well as to the "quasi producers," those who present themselves as producers by affixing their name, trademark, and other distinguishing feature on the product. Losses in the area of consumer goods are more likely; producers of industrial products (i.e., machines) will encounter more actions resulting under worker liability claims and compensation. Component producers may also face claims caused by the defective product components of third-party manufacturers. Figure 2-3 describes the product liability terms.

The definition of *supplier* has been broadened and now includes retail traders who may be subject to the secondary liability of the supplier. Where several persons are liable for the same damage, the injured person should be able to claim full compensation for the damage from any one of them. A major increase in the liability of the so-called quasi producers is now possible. These companies, which do not traditionally produce products but only select or distribute products manufactured by others, may be drawn into liability suits and/or corrective measure actions. These quasi producers include, but are not limited to, importers, suppliers, distributors, mail order houses, supermarkets, and department stores. The costs of litigation per defective product will be multiplied because of increasing complexity and multi-defendant litigation will also rise.

a. *Product* means all movables, with the exception of primary agricultural products (e.g., of the soil, fisheries) and game, even though incorporated into an immovable. *Product* includes electricity.

b. *Producer* means:

—the manufacturer of a finished product, the producer of any raw material or the manufacturer of a component part and any person who, by putting his name, trademark, or other distinguishing feature on the product presents himself as its producer;

—any person who imports a product for sale, hire, leasing, or any form of distribution in the course of his business shall be deemed a producer and shall be responsible as a *producer*; and

—where the producer of the product cannot be identified, each supplier of the product shall be treated as its producer unless he informs the injured person, within a reasonable period of time, of the identity of the producer or of the person who supplied him with the product. The same shall apply, in the case of the imported product if this product does not indicate the identity of the importer referred to above, even if the name of the product is indicated.

c. *Damage* means:

—damage caused by death or personal injuries; and

—damage to, or destruction of, any item of property other than the defective product itself, with a lower threshold of 500 ECU, provided that the item of property: (i) is of a type ordinarily intended for private use or consumption, and (ii) was used by the injured person mainly for his own private use or consumption.

d. *Defective product* is when a product does not provide the safety which a person is entitled to expect, taking all circumstances into account, including:

—the presentation of the product;

—the use to which it could reasonably be expected that the product would be put; and

—the time when the product was put into circulation.

FIGURE 2-3   Product Liability Terms

The Law now allows consumers to initiate civil actions independent of the authorities and removes the need for consumers to prove negligence in a legal action.

The consumer may take action against all parties involved in the supply chain simultaneously (joint and several). The manufacturer's best defense is likely to be that they took all available steps to ensure conformity (due diligence).

Directive 85/374/EEC concerning Liability for Defective Products (Product Liability Directive) was enacted in 1987 to protect consumers from defective products. This directive states that all producers involved in the production process shall be made liable, insofar as the finished "product, component, or raw material" supplied by them was defective and that the liability should extend to importers and persons who present themselves as producers by affixing their name to the product. To protect the physical well-being and property of the consumer, the defective nature of the product should be determined by reference not to its fitness for use but to the lack of safety that the public at large is entitled to expect: "This Directive therefore puts responsibility on the supplier to produce safe products, through the pressure which the costs liability places on him after an accident due to a defective product" (*OJEC*; 89/C 267/03).

A product is considered "defective" when it does not provide the safety that a person is entitled to expect.

The directive does not set financial ceilings on a producer's liability. The EU believes such limits inappropriate, but it recognizes that member states may deviate from that position because of differing traditions. The directive lets member states limit liability, but only if the limit is high enough to guarantee adequate protection for the user. Figure 2-4 shows a comparison between German and European liability laws.

The producer is liable for damage caused by a defect in the product. The injured person only needs to prove the damage, the defect, and the casual relationship between the defect and the damage.

To achieve effective protection of consumers no contracts are allowed to reduce the producer's liability. The liability of the producer arising from this directive may not, in relation to the injured person, be limited or excluded by a provision limiting his liability or exempting him from liability: "No contractual derogation (i.e., between producer and user) should be permitted as regards the liability of the producer in relation to the injured person."

In a few cases producers should be able to free themselves from liability if they furnish proof of certain exonerating circumstances. The contributory negligence of the injured person may be taken into account to reduce or disallow such liability. Producers shall not be held liable if they prove:

- That they did not put the product into circulation; or
- That it is probable that the defect did not exist at the time or that this defect came into being afterward; or
- That the product was neither manufactured by them for sale or any form of distribution; or
- That the defect was due to compliance with mandatory regulations issued by public authorities; or

Criteria	Germany cl 823 of BGB Fault Liability	EC Directive 85/374/EEC Liability
Conditions of evidence from plaintiff	Evidence of damage Product fault Cause of damage and product defect	Evidence of damage Product fault Cause of damage and product defect
Liability	Bodily damage Property damage Pain and suffering	Bodily damage Damage to privately owned property Maybe pain and suffering
Amount	Unlimited	Unlimited, possibly up to 80 million EURO EURO 560 deductible (approx.)
Liability time limit	30 years after product is first marketed	10 years after product is marketed
Cause of liability of defective product	Design fault Fabrication fault Instruction fault Negligence during production monitoring	Design fault Fabrication fault Instruction fault Possibly development fault
Liability chain	Manufacturer of end-product Component and material manufacturer Importer of technical equipment As part of their mandated liability insurance	Manufacturer of end-product Assembler Component and material manufacturer Third-country importer Retail merchant Quasi-manufacturer Fiction of the inclusion of the task range of an OEM supplier as if the product was manufactured by the OEM supplier.

FIGURE 2-4 Liability for Defective Products in Germany and Europe.
Courtesy of R&L Ing. Cons. GmbH, 65479 Raunheim, Germany.

- That the state of technical knowledge at the time they put the product into circulation was not such as to enable the existence of the defect to be discovered; or
- In the case of a manufacturer of a "component," that the defect is attributable to the design of the product in which the component was fitted or to the instructions given by the manufacturer of the product.

It is not reasonable to make the producer liable for an unlimited period for the defectiveness of a product; therefore, liability should expire after a reasonable length of time, without prejudice to pending claims. Member states shall provide a limitation of three years for the recovery of damages, which begins to run from the day on which the plaintiff became aware, or should reasonably have become aware, of the damage, of the defect, and of the identity of the producer.

As shown in this section there are only a few defenses for the manufacturer, such as it did not intend to market the product in Europe or exonerating circumstances. The producer will, however, be liable for "defective products" that cause "damage," especially when it has a "view to placing it on the market" and signify such by "affixing the CE marking."

💣☀ *Producers beware!* Placing the CE marking on a product informs users, competitors, and authorities that the producer is aware of the laws, standards, and consequences when a product becomes suspect.

Note. This section focuses on key elements of the directive and is only a part. Refer to the directives for the complete text. The actual wording of the directives is binding.

Customs Authorities and Border Controls

Directive 93/339/EEC on Checks for Conformity with the Rules on Product Safety of Imports states that "products may not be placed on the Community market unless they conform to the rules applicable; whereas member states are, thus, responsible for carrying out checks on their conformity." Furthermore, customs authorities must be closely involved in the market-monitoring operations provided for under Community and national rules. Customs authorities shall suspend the release of goods or batch and immediately inform the national authorities, who shall take suitable action. The conditions for suspending products release from customs and notification of the national authorities are:

- If goods give rise to a serious doubt as to the existence of serious and immediate risks;
- If customs authorities find that documentation that should accompany the product is missing; or
- If products are not marked as specified in the Community or national rules on product safety.

The following definitions are set down for the purpose of this directive:

- *National authority responsible for monitoring the market* means the national authority designated by the member state and required by it to check the conformity of product placed on the Community or national market with the applicable Community or national legislation;

- *Accompanying document* is any document that must accompany a product when it is placed on the market, in accordance with the legislation;
- *Marking* is any marking or labeling that products must bear in accordance with the legislation and that certifies that the product conforms with that legislation; and
- *Customs authority* shall mean the authorities responsible for the application of customs legislation.

If the national authorities find that a product presents a serious and immediate risk they shall prohibit the product from being placed on the market and ask the customs authorities to include the following endorsement on the commercial invoice accompanying the product and any other relevant document: "Dangerous product—release for free circulation not authorized—Regulation (EEC) No 339/93."

Where national authorities determine that the product does not conform with the Community or national rules on product safety, they shall take the appropriate measures, which may, if necessary, include prohibiting the product from being placed on the market and shall also ask the customs authorities to include the following endorsement on the commercial invoice accompanying the product and any other relevant document: "Product not in conformity—release for free circulation not authorized—Regulation (EEC) No 339/93."

To implement this regulation, Directive 81/1468/EEC was published on mutual assistance between the administrative authorities of the member states and cooperation between the latter and the Commission to ensure the correct application of the law on customs matters.

Note. This section focuses on key elements of the directive and is only a part. Refer to the directives for the complete text. The actual wording of the directives is binding.

Conformity Assessment Procedures: The Modules

Directive 93/465/EEC concerns the various phases of the "conformity assessment procedures," known as the *modules,* and the "rules for affixing the CE conformity marking," which are to be used in conjunction with the technical harmonization directives. This section focuses on the modules for the LVD, EMC, and Machinery directives and gives an introduction to the CE marking. This directive introduces a common doctrine of conformity assessment and methodology to facilitate the "placing on the market" of products. The methods should ensure full conformity with the Essential Requirements of the directives, in particular, for the health and safety of users and consumers. In addition, the directive clarifies the manufacturer's obligations for assessment procedures and the CE marking. The CE marking is used by enforcement authorities to facilitate controls on the market. Products may be placed on the market only after conformity assessment according to the applicable requirements (directives, standards, etc.) and after the manufacturer has affixed the CE marking (see Figure 2-5).

- The objective of a conformity assessment procedure is to give users, consumers, and authorities assurance that products conform to the various requirements as expressed in the directives;
- Conformity assessment can be subdivided into modules to control the design phase or the production phase; in certain cases these two functions are inseparable (e.g., modules A, G, and H);
- A product should undergo a control in both phases before being placed on the market if the results are positive;
- The directives shall set out the range of possible choices which can be considered which may be considered by the Council to give the authorities the acceptable level of safety they seek for a given product or product sector;
- In setting the range of choices open to the manufacturer, the directives will take into consideration, such issues as the appropriateness of the modules to the type of products, the nature of risks involved, the existence or non-existence of third parties, the types and importance of production, etc.;
- The directives set out the conditions in which a the manufacturer makes his choice as to the most appropriate modules for his production;
- The directives should attempt to leave as wide a choice to the manufacturer in setting the range of possible choices and avoid imposing unnecessary modules which would be too onerous;
- Notified bodies should apply the modules without undue burden on the economic operators in order to ensure consistent interpretation and application of the modules; the EOTC or Commission will organize close cooperation between the notified bodies;
- Wherever directives allow use of quality assurance techniques for manufacturers, they must also wherever possible provide for the possibility of recourse to product certification;*
- Member states shall notify only bodies, for operating the various modules, which comply with the requirements of the directives and accredited to EN 45000 series;
- The Commission shall publish lists of notified bodies in the *OJEC* and it shall be updated; and
- The CE marking (accompanied, where appropriate, by the notified body identification symbol) shall be affixed to show that the production phase has been carried out satisfactorily with regard to the requirements of the directives.

---

* Quality assurance techniques (e.g., ISO 9000 factory quality), are not relevant for modules A, Aa, or B-C and may not take the place of the safety/EMC directives for product conformity.

FIGURE 2-5   General Guidelines for Use of Conformity Assessment Procedures

📖   The aim of the CE marking is to "symbolize" the conformity of a product imposed by the directives and to "indicate" that the manufacturer/supplier has undergone all the evaluation procedures laid down by Community law. The CE marking is used to facilitate controls on the market by inspectors.

The procedures for conformity assessment are chosen from the various modules and in accordance with the criteria set out in the primary directives. Departure from the procedures is allowed only when the specific circumstances of a directive so warrant. Quality is shown in some of the modules, but a common mistake for some ISO 9000 quality–certified companies is to assume that they can select a module that utilizes ISO 9000 when this is not possible unless the primary directive allows it. There is no mention of ISO 9000 quality in the LVD, EMC, and Machinery directives (see Product Quality versus Factory Quality in Chapter 5). ISO 9000 quality may be necessary for some products (i.e., medical products) under the Medical Device Directive (MDD).

☞ ISO 9000 Quality is for companies and factories, it does *not* cover product safety or EMC conformity!

For most products and machines, the self-declaration process (module A) is possible. In practice the manufacturer performs the complete product assessment according to EU standards, issues the declaration, and affixes the CE marking to the product. A technical file or documentation must also be available on demand for national enforcement authorities. Keep in mind that this is an internal self-assessment process, a "do-it-yourself" approach, that results in issuance of the manufacturer's declaration of conformity and the CE marking (Figure 2-6). The buyer may demand proof of safety/EMC compliance in the form of a mark, certificate, or test report from a European notified or competent body.

All products covered by the directives must be assessed for conformity and bear the CE marking prior to entry into the European Union. The quantity of products or machines or the nature of the transfer does not relieve the manufacturer from performing the conformity assessment and from affixing the CE marking. This applies to all products and machines put into service in a public or private capacity, for professional or nonprofessional use, and "for payment or free of charge." In certain circumstances, such as trade shows, products may be allowed into the European Union for a limited time with the proper warnings on the product. Such units may be operated only by the manufacturer.

The essential objective of a conformity assessment procedure is to enable the authorities to ensure that products placed on the market conform to the requirements expressed in the directives; and, these products must meet the "high level of safety" that the directives seek for a given product or product sector. There are seven modules in 93/465/EEC, and the generic directives (LVD, EMC, etc.) or product-specific directives (machinery, etc.) will set the range of choices that may be considered, as follows:

> *Self-declaration.* This route is available for products/machines where a mandatory type exam is not required. This route allows "internal control of production" through product assessment and testing by the manufacturer. Under self-declaration, the manufacturer takes complete responsibility for the assessment, testing, documentation, and declaration of conformity and CE marking.

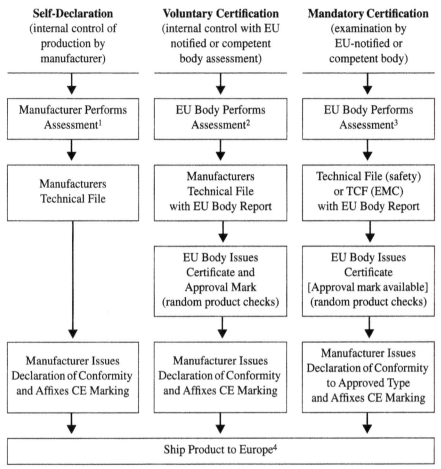

Self-Declaration (internal control of production by manufacturer)	Voluntary Certification (internal control with EU notified or competent body assessment)	Mandatory Certification (examination by EU-notified or competent body)
Manufacturer Performs Assessment[1]	EU Body Performs Assessment[2]	EU Body Performs Assessment[3]
Manufacturers Technical File	Manufacturers Technical File with EU Body Report	Technical File (safety) or TCF (EMC) with EU Body Report
	EU Body Issues Certificate and Approval Mark (random product checks)	EU Body Issues Certificate [Approval mark available] (random product checks)
Manufacturer Issues Declaration of Conformity and Affixes CE Marking	Manufacturer Issues Declaration of Conformity and Affixes CE Marking	Manufacturer Issues Declaration of Conformity to Approved Type and Affixes CE Marking

Ship Product to Europe[4]

[1] Manufacturer's assessment of the directive on the basis of conformity with European harmonized standards and/or the ERs.

[2] EC body assessment based on European harmonized standards or other criteria, that is, when no harmonized or equivalent national standard exists or in the case of innovation, another specification deemed acceptable by EC body.

[3] EC body assessment and approval of technical file based on European harmonized standards and/or other criteria (i.e., Annex IV listed machinery (89/392/EEC) or for EMC where standards are only applied in part).

[4] A technical file, declaration of conformity and CE marking is required under all three procedures.

Figure 2-6 Conformity Assessment Procedures

Voluntary certification. Oftentimes manufacturers have products assessed by a European body (notified or competent) for a "mark and certification" for marketing purposes and as a defense of "due diligence" in the event of a challenge. The "intervention of notified body" certification route provides confirmation of accuracy of the testing and documentation (test reports, etc.).

Mandatory certification. Most products and machines do not require mandatory certification. But in such cases, as with some high-risk machinery or when harmonized standards do not exist or are not applied in full, a "type examination" by a notified body is required. After successful testing, a *Type-Exam Certificate* for machinery or *Certificate of Conformity* for EMC is issued by the EU Body. The manufacturer then affixes the CE marking and issues a declaration of conformity.

Note. In all three cases, the CE marking, declaration, and technical file shall be in place. ISO 9000 factory quality techniques are not a part of modules A, Aa or B-C (see Product Quality versus Factory Quality in Chapter 5). The relevant safety/EMC directives and conformity assessment modules still apply.

Technical Files, Documentation, and Test Reports

With Europe's new American-style liability laws, preventing accidents through the proper use of warnings and design documentation gains more importance. The number and size of liability claims against producers, distributors, and importers are increasing. The manufacturer must be able to prove that it considered all relevant safety/EMC aspects and supplied a product without defects. Because a "no-fault" defense is no longer viable, the manufacturer's ability to prove that the product was not "defective" at the time it was put into circulation gains even more significance. Simply showing that a certain model or serial number was manufactured is not enough. The product suppliers must be able to prove, through their documentation, that adequate design, production, and quality control procedures were in place to ensure a compliant product, that is, without defects. The supplier and their network of distributors now must also retain the documentation for a longer time. The documentation must be retained for 10 years after the last production but may be required beyond this date if losses from a specific product occurred. Without the documentation the suppliers will hardly be in a favorable position to defend themselves, even if there were no defects. A "technical file" includes the documentation that proves that a product conforms to relevant directives and standards.

The technical file is for market surveillance purposes and must be kept at the disposal of national enforcement authorities. The documentation must be readily available to a *duly substantiated request* by enforcement authorities for inspection and control purposes. Failure to make the documentation available may constitute sufficient grounds for doubting the presumption of conformity. It is the manufacturers or their authorized representative in Europe who are ultimately responsible for the accuracy of the technical file. The technical files' accuracy and the product safety/EMC conformity are of paramount importance.

📖 Technical files are for market control and must be readily available for the enforcement authorities.

The *technical file* consists of the technical documentation necessary to demonstrate the conformity of the product to the essential requirements of the directives (LVD, EMC, Machinery, etc.). It shall cover the design, manufacture, and operation of the product. The file may be in English or another EU language and should only address the safety/EMC issues, in other words, kept to a minimum. The contents of the file depend on the applicable directive, but in general consists of the following:

- Declaration of conformity;
- Name and address of the manufacturer;
- General description and identification of the product;
- List of harmonized standards applied;
- Solutions adopted to satisfy essential requirements and rationale;
- Examination results and calculations;
- Test reports, test data, components lists, certificates, etc.;
- Design drawings, circuit diagrams, etc.;
- Operation manual, warnings, etc.; and
- Measures adopted to ensure ongoing compliance.

Test reports are technical records on the conformity *assessment* of a product according to specific standards. Test reports are concise accounts, including clause-by-clause details on the results of the product assessment, standards rationale, test data, construction, and components. Test reports are an essential tool for conformity assessment and *the* most important element of the technical file. The manufacturer may obtain a test report as a element of proof of conformity from an notified/competent body (not mandatory) in advance and keep it with the technical file. "The availability of such a report [from a notified body] would make matters easier and speedier in the event of challenge by authorities" (refer to Notified Bodies and Due Diligence sections in Chapter 4).

For safety assessments, the testing performed may actually be minimal depending on the equipment in question and the standard(s) applied. The safety assessment usually involves considerable time verifying a product's conformity with the construction and components requirements according to all the clauses of the applicable standards (see Design Guidance in Chapter 6). For EMC the majority of the assessment and time spent involves the tests themselves. The resulting EMC report(s) contain, for the most part test data and test configuration, with some construction and components information. If the test report is not available, the customer and/or certification body has the right to doubt conformity of the component, product, or machine under consideration.

📖　　Test reports may be requested by customers, certification bodies, or enforcement authorities for review and verification purposes.

The EMC Directive sometimes mandates the use of a competent body for mandatory certification and a special file called the *technical construction file (TCF)*. This is the case when no harmonized standards exist or they are applied only in part. The TCF is generated by the manufacturer in conjunction with the competent body. This process is commonly called the "TCF route" and is a certification process that requires a "Certificate of Conformity" issued by a competent body prior to CE marking. Under the TCF route, the manufacturer may place the CE marking on the product only after satisfactory testing, completion of the TCF, and receipt of the "Certificate of Conformity" from the competent body. The TCF process is often times used when there are numerous product variations or for large machines.

Declaration of Conformity

The EU declaration of conformity is the procedure whereby the manufacturer or authorized representative "ensures and declares" that the products concerned satisfy the requirements of the directives that apply to them. The declaration of conformity shall be signed by the manufacturer, wherever located, or by the authorized representative established within the Community. The LVD and EMC Directives state that the declaration needs to be kept on file, but manufacturers may also supply a declaration with each product or shipment. For machinery, a declaration must accompany each machine. The declaration shall be in the same language as the original instructions for use. The declaration of conformity shall contain the name and address of manufacturer or representative; description of product (name, model number, etc.); directive(s) declared; harmonized standards applied; additional standards and specifications (where appropriate); place and date of issue; and name and signature of authorized person. The format is not important as long as the declaration contains the necessary elements laid down in the appropriate directives. The declaration should be drawn up in the same language as the original instructions of use and must be typewritten or handwritten in block capitals. The following declarations (Figure 2-7) are for illustration only and not a recommendation. Standards and requirements may be changed or added at any time.

CE Marking Guidelines

According to the *Official Journal,*

A single CE marking should be used in order to facilitate controls on the Community market by inspectors and clarify the obligations of economic operators [manufacturer or supplier] . . . ; the aim of the CE marking is to symbolize the conformity of a product . . . and to indicate that the economic operator has undergone all the evaluation procedures laid down by Community law in respect of his product. (*OJEC*; 93/465/EEC)

It is well advised that manufacturers and suppliers be informed not only of the CE marking's advantages, but also of its limitations:

- The CE marking is *not* a mark or certification or approval.
- The CE marking is *not* for sales or marketing or promotion.

Declaration of Conformity

ABC GmbH
Street
City, Postal Code
Germany

Type of Equipment	**Transparency Projector**
Model Name	**Newview+**
Model/Type Number	**NV1ZXYZ**
Year of Manufacture	**1995**
Application of Council Directives	**73/23/EEC and 89/336/EEC**
Standards to Which Conformity Is Declared	
Safety:	**EN 60335-2-56:1991,**
EMC:	**EN 55014:1993,**
	EN 60555-2/-3:1987,
	EN 50082-1:1992
Manufacturer's Name	**ABC GmbH**
Manufacturer's Address	**Street**
	City, Postal Code
Importer's Name	**XYZ Imports Co.**
Importer's Address	**Street**
	City, Germany

We hereby declare that the product specified above conforms to the above-mentioned directives and standards.

(Notice: *This Declaration of Conformity is supported by the attached safety/EMC approval certifications, numbers S9512345 and V9554321 issued by TUV Rheinland, a European certification body.*)

Signature

(name, title, location, date)

FIGURE 2-7a   Sample Declaration: Overhead Projector

- The CE marking is *not* a quality marking.
- The CE marking is *not* for components (some exceptions, see Components in Design Guidance in Chapter 6).
- The CE marking is *based on compliance with EU directives (ERs) and standards (ENs)* (use of non-European requirements or standards may require the involvement of an EU-notified body).

Declaration of Conformity

Manufacturer's Name:	ABC Co.
Manufacturer's Address	ABC Street
	Detroit, MI 90061 USA

declares that the product:

Product Name:	Laser Printer
Model Number:	5432X

conforms to the following standards:

Safety: EN 60950,
EN 60825

EMC: EN 55022, Class B,
EN 50082-1, using:
EN 61000-4-2 (IEC 801-2),
EN 61000-4-3 (IEC 801-3),
EN 61000-4-4 (IEC 801-4)

The product is in conformity with the requirements of the Low-Voltage Directive (73/23/EEC) and the EMC Directive (89/336/EEC).

European Contact:
John Doe, Quality Manager
ABC Co., Bonn, Germany
(Fax: 49-XXXX-XXXXXX)

Signature

(name, title, location, date)

FIGURE 2-7b   Sample Declaration: Laser Printer

The term *CE mark* was changed to *CE Marking* in the New Approach Directives. The CE marking is *not* a mark and therefore must not be confused with a "mark, certificate, or approval" issued by an accredited certification body, as listed in the *Official Journal*. Rather, the CE marking is a "symbol" of the manufacturer's declaration of conformity that implies conformity with the "minimum requirements" set out in the Directives. The CE symbol is *not* a registered mark, which is in principle under the manufacturer's own responsibility. The CE marking is a declaration for the inspectors (i.e., customs) and allows the product to be "placed on the market."

Declaration of Conformity

Manufacturer:	ABC Co. 1234 Street San Francisco, CA USA
Factory:	ABC Co. 1234 Street San Francisco, CA USA
Name of representative:	XYZ Co. Address EU COUNTRY
Machinery description:	Semiconductor Pick and Place Machine.
Model/Type number:	IC/XXX-YY

I hereby declare that the machinery complies with the Essential Health and Safety Require-ments of the Machinery Directive (89/392/EEC) and with the provisions of the Electromag-netic Compatibility Directive (89/336/EEC), as amended by 93/68/EEC.

Safety Standards:

EN 292-1/-2:1991	Safety of Machinery—Basic concepts, general principals for design
EN 60204-1:1993	Safety of Machinery—Electrical equipment of machines—General requirement's
EN 1050:1996	Safety of Machinery—Risk assessment
EN 1088:1996	Safety of Machinery—Interlocking devices associated with guards
EN 954-1:1992	Safety of Machinery—Safety related parts of control systems—General principals
EN 614-1:1995	Safety of Machinery—Ergonomic design principles
EN 418:1992	Safety of Machinery—Emergency stop equipment, functional aspects
EN 349:1993	Safety of Machinery—Minimum gaps to avoid crushing of parts of the human body
EN 294:1992	Safety of Machinery—Safety distances to prevent danger to upper limbs

Additional Safety Standards:

EN 60947-5-1:1991	Low-Voltage switch gear & control gear, electromechanical control cir-cuit device
EN 60825-1:1994	Safety of Laser Products—Equipment classification, requirements and user guides
EN 60447:1994	Man–machine interface (MMI)—Actuating principles

EMC Standards:

EN 55011:1991	Limits and Measurement of RFI for Industrial, Scientific, and Medical Equipment
EN 55022:1994	Limits and Measurement of RFI for Information Technology Equipment
EN 50082-2:1995	Generic Immunity for Heavy Industrial Environment

Signature
_____

(name, title, location, date)

Note: If or when a machine safety standard (type C) is published for this machine type, some of the safety standards are then superseded by it.

FIGURE 2-7c Sample Declaration: Machine

📖 There is no marketing advantage with only the CE marking since *all products will have the CE marking.* Products with the CE marking are therefore assumed equal by customs inspectors, but customers may expect more.

The CE marking is affixed at the end of the production phase by the manufacturer. In exceptional and duly warranted cases, the manufacturer's authorized representative or the person responsible for placing the product on the market may be allowed to affix the CE marking. The CE marking must be affixed to the product itself. In certain cases the Directive does allow the CE marking on the packaging, such as when the product's size is too small. Other conformity "marks" indicating compliance with European standards are allowed, provided they are not liable to deceive third parties as to the meaning of the CE marking and provided such marks do not cause confusion with and do not reduce the visibility of the CE marking. Figure 2-8 shows the general guidelines for the CE marking.

Written questions concerning the CE marking were posed by a member of the European Parliament to the Commission. The answers were given by Vice President Mr. Bangemann on behalf of the European Commission and published in the *Official Journal* (*OJEC*; 95/C 326/50). The first five questions and answers are reproduced here:

Q.1 Does the EC verification mark on a product guarantee free access to the whole internal market for the product concerned?

A.1 Yes. The CE marking can be described as a "passport for industrial products" allowing them to circulate freely throughout the European Economic Area (EEA). It is a mandatory conformity marking which shows the compliance of products with all provisions of 16 Directives which relate to safety, public health, consumer protection, or other essential requirements of community interest.

Q.2 Is the EC verification mark recognized in all countries of the European Union?

A.2 Yes. It addresses the market surveillance authorities of the member states and aims to facilitate their surveillance tasks by visibly demonstrating conformity. Of course, where a member state ascertains that products bearing the marking do not comply with the requirements of the directives applicable, it takes appropriate measures to withdraw the products from the market, to prohibit the placing on the market and putting into service, and to restrict free movement.

Q.3 Is the Commission aware that, in some countries in the Union, other seals of approval are still in use in addition to the EC verification mark?

CE Guidelines:

- Symbolizes conformity to all obligations on the manufacturer by virtue of the directives and in specific cases such conformity may not be limited to the essential requirements;
- Symbolizes that the person affixing the CE marking has verified that the product conforms to all the Community provisions and has been subjected to the appropriate evaluation procedures;
- Where products are subject to other directives the CE marking must indicate that the products are also presumed to conform to the provisions of those other directives;
- The CE marking must consist of the initials "CE" taking the appropriate form and must have a height of at least 5 mm;
- The CE marking must be affixed to the product or its data plate, however when this is not possible or not warranted it must then be affixed to packaging and to the accompanying documents;
- Any product covered by the directives must bear the CE marking except where specific directives provide otherwise;
- The CE marking shall be affixed at the end of the production control phase;
- When a notified body is required and involved in the assessment (i.e., type exam), the CE marking must be followed by the identification number of the notified body;
- The affixing of any other marking liable to deceive third parties as to the meaning of the CE marking is prohibited;
- A product may bear different marks, for example marks indicating conformity to national or European "standards," provided such marks are not liable to cause confusion with the CE marking;
- The CE marking must be affixed by the manufacturer or their agent within the Community;
- Member states must take all provisions of national law to exclude any possibility of confusion and prevent abuse of the CE marking; and
- Where a member state establishes that the CE marking has been affixed unduly, the manufacturer, agent, or supplier is obliged to make the product comply and end the infringement under conditions imposed by the state and where noncompliance continues the member state must restrict, prohibit, or withdraw the product(s) from the market.

What the CE marking is:

- Is a manufacturer's or agent's self-declaration (a "do-it-yourself" approach);
- Is a symbol of a products conformity to the essential requirements of the directives;
- Allows products to be "placed on the market" and "ensure free movement of goods";
- Indicates manufacturer has undergone all the relevant procedures;
- "Presumption of conformity" is assumed for customs inspectors only when products are in conformity with European harmonized standards (ENs);
- Is for inspectors checks to facilitate controls on the Community market; and
- Is for "finished products" such as equipment, machinery, systems, apparatus, and devices that are complete and ready for use, by the operator, as a single commercial unit.

What the CE marking is not:

- Not for sales, marketing, or promotion;
- Not a mark, certification, or approval from European recognized "third-party" (e.g., notified body);
- Not a quality marking to influence consumers or users (modules A, Aa, B-C); and
- Not for components, since components have no autonomous use (except in a few cases, see Components in Design Guidance in Chapter 6).

FIGURE 2-8 General Guidelines for Affixing and Use of the CE Marking

A.3 Yes. According to Council Decision 93/465/EEC of 22 July 1993 concerning the rules for affixing the CE conformity marking, "a product may bear different marks." Other product marks are only prohibited if they are "liable to deceive third parties as to the meaning and form of the CE marking" or if they are "liable to cause confusion with the CE marking." Quality markings, as opposed to the CE marking, are voluntary, address consumers or users, and tend to influence their appreciation toward the relevant product. Thus, they have a different function to that of the CE marking. They are therefore acceptable.

Q.4 Is the Commission aware that such national seals of approval are held in higher esteem?

A.4 Yes. The CE marking is not a quality marking although it is often wrongly perceived as such and then compared to other quality marks. The Commission tries to remedy this situation by informing industry as well as the authorities responsible for market surveillance about the correct meaning and function of the marking. Furthermore, the second version of the "guide to the implementation of Community harmonization Directives based on the new approach and global approach" will include a chapter on the marking, clarifying its role.

Q.5 Does this development constitute a real danger to the internal market, bearing in mind that the EC marking is intended to eliminate technical barriers to trade?

A.5 No. It is true that the Community harmonization Directives which provide for the affixing of the marking aimed to remove technical barriers to trade and has as their objective the establishment and functioning of the single market.

However, the free circulation of goods which is assured by the marking does not necessarily mean that people will buy the products. In order to market and sell a product successfully a manufacturer often has to do more than what is required by legislation.

CHAPTER 3

Standards:
The Technical Rules

Better to ask twice than lose your way once.

DANISH PROVERB

Harmonized Standards: Presumption of Conformity

Standards in Europe date back to the late 1800s. The International Electrotechnical Commission (IEC) was founded in 1906 to establish safety specifications for energy and electrical products and protect consumers and the environment. IEC standards are prepared with consultation from technical committees (TC) in over forty countries. These committees have representation from many interested parties, including, but not limited to, manufacturers, authorities, notified bodies, and consumer groups. Another objective of the IEC is to provide a common reference for international trade through standards. This goal of one set of universally recognized standards is fast becoming a reality for worldwide product acceptance.

The concept of European Conformity (CE marking) revolves around European harmonized standards as the *minimum requirements* for product design and assessment. Strict adherence to these technical specifications should be the focus of all designers and manufacturers. Manufacturers can and should do more than the standards require and certainly should not do less. In reference to the conformity assessment modules the CE Marking Directive 93/68/EEC (*OJEC*; L220/21) states: "the choice of procedures must not lead to a lowering of safety standards of electrical equipment, which have already been established throughout the Community."

The European Commission mandates the creation of standards to support the essential requirements (ERs) of the directives. A standard is considered harmonized at the time of announcement which is published in the *Official Journal of the European Communities (OJEC)*. Compliance with the harmonized standards will, in most cases, ensure a product's conformity with the essential requirements of the directives. Adherence to European harmonized standards is the only proven and universally accepted method of showing conformity with the ERs of the directives (Figure 3-1). Properly applying standards produces a "presumption of conformity."

✓ A "presumption of conformity" is conferred by the use of harmonized standards. Products will *not* benefit from a presumption of conformity when harmonized standards are not properly applied.

European standards (norms), are usually based on IEC standards. For example EMC standard EN 61000-4-2 (Immunity to Electrostatic Discharge) is based on IEC 1000-4-2, and product safety standard EN 60335 (Safety of Household Appliances) is based on IEC 335. All European standards (ENs) go through rigorous review by any and all interested parties. National authorities, whose primary responsibility is public safety, participate in the review process.

📖 National enforcement authorities are obliged to recognize that products in conformity with harmonized standards ("EN" listed in *OJEC*) and transposed into national standards are presumed to conform to the ERs of the directives.

When a product becomes suspect the national enforcement authority will typically use the EN or equivalent national standards (usually with the help of a notified body) to evaluate the product's conformity or lack thereof. Taking any route other than the harmonized standards approach may work against those who do not conform with the recognized standards, since compliance with these standards is universally recognized and expected by the enforcement authorities, notified bodies, and customers.

Meeting the rules set out in the European standards is the *minimum* criterion, and most manufacturers should not only meet but also endeavor to *exceed* these requirements. Even when the product conforms to the relevant standards, the authorities are not barred from taking the appropriate actions against manufacturers where there is evidence that a product is dangerous (see General Product Safety in Chapter 2).

The Low-Voltage Directive (LVD) 73/23/EEC
• "Products are presumed to conform to these objectives [ERs] where the equipment has been manufactured in accordance with technical standards which in their order laid down in the directive, are as follows: — harmonized standards, drawn up in accordance with Article 5 . . . — where harmonized standards . . . have not been drawn up and published, international rules [IEC, etc.] . . . — where harmonized standards . . . or safety provisions . . . are not yet in existence, the [national] standards in force in the Member State of manufacture (Article 7)" (C59, p. 3). • "Whereas Decision 90/683/EEC (*) establishes the modules for the various phases of the conformity assessment procedures which are intended to be used in the technical harmonization directives; whereas the choice of procedures must not lead to a lowering of safety standards of electrical equipment, which have already been established throughout the Community" (93/68/EEC, No L220, p. 21). • "The EC declaration of conformity must contain the following elements: reference to the harmonized standards" (see directive for other elements) (93/68/EEC, No L220, p. 21).

The Machinery Directive 89/392/EEC
• "Where a national standard transposing a harmonized standard, the reference for which has been published in the *Official Journal of the European Communities,* covers one or more of the essential safety requirements, machinery or safety components constructed in accordance with the standard shall be presumed to comply with the relevant essential requirements." • "In absence of harmonized standards, member states shall take any steps they deem necessary to bring to the attention of the parties concerned the existing national standards and specifications that are regarded as important or relevant to the proper implementation of the essential safety and health requirements." • "Contents of the EC declaration of conformity for machinery (1): — where appropriate, a reference to the harmonized standards, — where appropriate, the national standards and specifications used, (see directive for other elements)."

The Electromagnetic Compatibility (EMC) Directive 89/336/EEC
• "To facilitate proof of conformity with these requirements, it is important to have harmonized standards at European level concerning electromagnetic compatibility, so that products complying with them may be assumed to comply with the protection requirements." • "Whereas, pending the adoption of harmonized standards . . . the free movement of goods should be facilitated by accepting . . . apparatus complying with the national standards." • "Member states shall presume compliance with the protection requirements referred to Article 4 in the case of apparatus which is in conformity: — with the relevant national standards transposing the harmonized standards . . . — or with the relevant national standard . . . in the areas covered by such standards, no harmonized standards exist."

Notes: The above quotes are taken word for word from the relevant EU Directives. The LVD, EMC, and Machinery Directives are amended by 93/68/EEC adding the CE Marking. Directive 90/683/EEC, mentioned in 73/23/EEC, is replaced by 93/465/EEC. © Lohbeck.

FIGURE 3-1 Directives Require Harmonized Standards for Presumption of Conformity

Standards Organizations		Primary Subject	Standards Examples
IEC (CISPR)	International Electro-technical Commission (EMC section of the IEC)	Electrical Safety (EMC)	IEC 65, 204, 335, 950, 1010 (CISPR 11, 14, 22)*
CENELEC	European Committee for Electrotechnical Standardization	Electrical Safety (EMC)	EN 60065, 60204, 60335, 60950, 61010 (EN 55011, 55014, 55022)
CEN	European Committee for Standardization	Mechanical and System Safety	EN 292, 349, 418, 954, 1050
ISO	International Organization for Standardization	general	ISO 9000, 9241, 14000

* Note: In 1997, the IEC introduced a new numbering system for all its international standards. Existing standards also adopt this new numbering system; for example, IEC 65 is now referred to as IEC 60065.

The first step in the development of a standard begins with the European Commission mandating that a standards organization draft a standard. Next, a closing date is established, and within this time period all interested parties may comment. After approval, the standard will be published as an harmonized standard in the *Official Journal,* and then it is transposed into national standards by the member states. A standards implementation example would be:

EN 55022:1994, Limits and Methods of Measurement of RFI Characteristics of ITE:

- DOR 12/09/1992   Date of ratification by technical board
- DOA 03/15/1993   Date by which EN may be announced at national level
- DOP 12/15/1994   Date when EN is transposed into national standard
- DOW 12/31/1995   Date when national standards conflicting with EN must be withdrawn

European standards are clearly the main path toward conformity. Although use of European standards is voluntary in one sense, European harmonized standards have become the technical governing rules and are in reality the *obligatory* way for manufacturers to reduce unnecessary risks and to meet the EU directives. In practice, only published European harmonized standards, such as ENs (EN = European Norm), should be used to show conformity. European harmonized standards (ENs) offer the simplest means of meeting the essential health and safety requirements (EHSRs) of the directives.

The Commission Act on the Global Approach states:

The Council resolution of 7 May 1985 showed the way by accepting that there could be more than one means of proof of conformity to a directive. It provided for presumption of conformity to a directive on the basis of a European harmonized standard, or, during a transition period, of national standards which have submitted and recognized as equivalent under Community control procedure. When the manufacturer complies with these standards, the directives allow him to make use of simplified certification mechanisms. When the product does not conform to a standard, however, either because the standards do not exist or because the manufacturer, for in the case of innovation, prefers to apply other manufacturing criteria of his own choice, the assessment of conformity to the essential requirements must [may] involve a third party either by certification or by third party testing [via notified/competent body]. (*OJEC*; 89/C267/03)

Furthermore, the General Product Safety Directive states:

Where there are no specific Community provisions [i.e., European harmonized standards] governing the safety of the products in question, a product shall be deemed safe when it conforms to the specific rules of national law of the Member State in whose territory the product is in circulation. (*OJEC*; 92/59/EEC, art. 4)

Standards are valid for conformity assessment after (1) publication in the *Official Journal* as a harmonized standard and (2) transposition into national standards. As an example, some of the official national versions of the ENs for Safety of Information Technology Equipment (ITE) are:

- German version: DIN EN 60950 = EN 60950 ≅ IEC 950
- British version: BS EN 60950 = EN 60950 ≅ IEC 950
- French version: NF EN 60950 = EN 60950 ≅ IEC 950

The harmonized standards are listed/appended to the directives themselves. The proper European standard must be applied for conformity assessment as follows, in the following order of preference:

1. A European harmonized standard (EN) referenced in the *Official Journal*.
2. When no harmonized standard (EN) exists, an IEC standard published in the relevant directive.
3. When no harmonized standard (EN or IEC) exists, a European *National Standard* (i.e., DIN, VBG, BS) in force in the member state.
4. In absence of standards mentioned above, Community specifications, member state standards, codes of practice in the sector concerned or to the state of the art and technology to the safety which the EU consumer may reasonably expect.

In the last two cases it is advisable (and sometimes mandatory) to consult a notified body for guidance and testing. Authorities are obliged to recognize that products manufactured in conformity with harmonized standards are presumed to conform to the essential requirements of the directives.

📖 Over 5,000 EN, 5,000 IEC, and 10,000 ISO standards have been issued so far!

Of the 5,000 EN standards issued, over 600 have been published in the *OJEC* as of this writing, with many more to come. Since most ENs are based on IEC standards, compliance with the EN standards also ensures conformity with equivalent IEC standards. The world is following Europe's lead, and most countries have, or eventually will have, accepted the European standards or the IEC equivalents. Therefore, meeting the European standards will also help manufacturers to comply with the technical standards worldwide.

Selecting Standards

Compliance with the EN standards is the *minimum acceptable criterion* for conformity, so I'll start by identifying the appropriate standards (also see Essential Requirements). The science of selecting the appropriate standards can be simple or complicated depending on your experience and product type. Exercise care when identifying the proper standards that may apply to your product design. The selection of a product standard may be a relatively simple task, as for many Low-Voltage Directive (LVD) products, or cumbersome, as for some machines. You must consider the product's intended usage and make sure the product falls within the standard's scope.

📖 The regulations demand the proper choice and application of standards. Variance from this procedure is not at the manufacturer's discretion.

When choosing standards, consider the following points:

1. *Type of product.* Check the scope of the standard to determine if it refers to the product.
2. *Environment.* Make sure the standard deals with conditions of usage.
3. *User.* The skill and protection of the operator and service person are important.
4. *Other considerations.* Other standards may be applicable such as for non-tested components, environment, guarding, state of the art, ERs of directives (see Essential Requirements below).

📖 Tunnel vision is commonplace for beginners. Assuming only one standard applies or applying the improper standards and not addressing all the ERs of the directives may lead to nonconformity.

Beware of the dangers of tunnel vision! Manufacturers may limit their view; machine builders may focus on the mechanical guarding issues, ignoring the electrical; electrical product manufacturers may address the electrical concerns, ignoring the mechanical. Manufacturers must address all hazards even when a standard does not mention a specific hazard (see Essential Requirements below). If an electrical product (LVD) has a potential hazard, such as mechanical or laser, that is not covered by the relevant electrical safety standard, then the designer must identify other standards to address these issues. In addition, if critical components (i.e., > 50 Vac/60 Vdc or in safety circuit) do not bear a Type-Approval Mark from a EU recognized body, then the designer must confirm that it meets the relevant EN/IEC standards or perform a complete evaluation and testing of the component according to the relevant component standards. Remember: A CE marking on a component is *not* an EU third-party mark; additional testing may be necessary (see Components in Chapter 6).

In general, only one standard is needed, but in some cases several standards are required, such as for electrical, mechanical, documentation, and others. Also, do not forget the ERs of the relevant directives and the market expectations (refer to Essential Requirements in this chapter and Product Liability in Chapter 2).

All hazards and relevant standards must be addressed, this means that no hazard or component is to go unchecked even if the main product standard does not address a particular hazard, risk, component, etc.

The standards and other regulations address all possible hazards concerning the protection of consumers, animals, property, and the environment. The operator of the product/machine is generally the focus of protection, but the standards may require protection of service personnel (see Design Guidance in Chapter 6).

Possible hazards are:

- Electrical
- Energy
- Mechanical
- Temperature
- Fire
- Chemical
- Gases
- Radiation
- Pressure
- Others

Standards, like directives, come in three forms: types A, B, and C, otherwise called *basic, generic,* and *product-specific* standards. The type C product-specific standards are the top-level standards and take precedence over types A and B standards. Refer to the appropriate standards as needed.

Basic standards (type A). These fundamental standards contain general principles for safe design or measurement techniques/levels for EMC and may be applied to products when appropriate. The A and B standards are especially important for EMC and machinery. Some examples of basic safety standards for machinery are EN 292-1/-2 (Design Concepts), and EN 1050 (Risk Assessment). The EN 61000-4-X series (IEC 801-X) for immunity levels is an example of basic standards for EMC.

Generic standards (type B). For safety, the type B standards concern specific technical aspects and are applied as needed, such as for specific components or guarding. For machinery the B standards are further divided into B1 and B2 standards.

B1 Standards. These apply to particular aspects, such as surface temperatures and safety distances. Some B1 examples are EN 418 (Emergency Stop Systems), EN 954-1 (Safety Related Control Systems), and EN 60204-1 (Electrical Requirements of Machinery).

B2 Standards. These apply to particular safety devices or components such as EN 574 (Two Hand Controls) and EN 60947 (Safety Switching Devices).

For EMC, the generic standards are grouped by function and environment, such as products for use in heavy industrial areas. The generic EMC standards list performance criteria and refer to type A standards for tests. Some examples of type B standards are EN 50082-1/-2 immunity requirements for Residential/Commercial/Light Industrial Areas (-1) and Heavy Industrial Areas (-2).

Product standards (type C). Type C standards address a specific product or related group or range of equipment. The type C standards take precedence over the basics and generics and call up the appropriate A and B standards, when necessary. The type C nomenclature varies slightly between the directives with the term *product-specific standard* being the most widely recognized for type C's:

- Low-Voltage Directive, product-specific standard
- Machinery Directive, machine safety standard
- Electromagnetic Compatibility Directive, product family (or specific) standard

There are numerous product standards. For electrical LVD products: EN 60950 (Safety of ITE), EN 61010-1 (Safety of Measurement and Lab Devices), EN 60335 (Safety of Household and Similar Appliances). For machinery: EN 415 (Safety of Packaging Machines), EN 201 (Safety of Plastics Molding Machines), EN 1010 (Safety of Paper Machines).

For EMC, Product Family Standards examples are: EN 55011 (Emissions for Industrial, Scientific, and Medical Products), EN 55022 (Emissions for Information Technology Equipment), EN 55014 (Emissions for Household Appliances), and EN 55104 (Immunity for Household Appliances).

Note. In some cases a type C product standard may not exist for a particular machine or product. In this case type A and B standards are used, along with the relevant EU national standards (if available), to assess the machine for safety or EMC, and these standards are then listed on the declaration of conformity for the CE marking. Use of a EU-notified/competent body may be required for products where no type C standard exists.

Essential Requirements and State of the Art

In addition to the standards the manufacturer must check the essential requirements of the directives and take into account the *generally acknowledged state of the art* for EMC and safety. Products, machinery, and components must be designed and manufactured in such a way that, when used under the conditions and intended purposes, they will not compromise the health and safety of persons or the environment. In selecting the most appropriate design solutions, the manufacturer shall apply the following principles in order of preference:

1. *Inherently safe design.* Eliminate or reduce risks as far as possible by construction.
2. *Guarding and safety components.* When a inherently safety design is not possible, use proper protective measures such as guarding, safety components, alarms, in relation to risks that cannot be eliminated.
3. *Warnings and instructions.* Inform users of residual risks due to shortcomings of design. This method is to be employed only as a last resort and when 1 or 2 is not possible.

☠ Warnings may not take the place of a safe design. If a product or machine can be made safe, it should be!

According to the ERs of the directives, compliance with the EN standards is defined as the *minimum* acceptable criteria, but in some cases the manufacturer will have to go *beyond* the standards, such as when:

- EN or IEC standard does not exist (shall use EU national standard until EN sanctioned);
- Existing standards are inadequate (preliminary, draft, or provisional standard may be recognized);
- "State of the art"—meeting the latest state of technology may be required in addition to standards;
- Justified national requirements or exceptions exist;
- Additional marking, documentation, or translation requirements;
- Ergonomics, noise, environmental protection, other;
- Mechanical in addition to electrical (for LVD products) or visa versa (for machinery);
- Special usage or environment (i.e., industrial area, hazardous location, EM fields);
- Products safety influenced by EMC (EMC tests may also be required for safety compliance);
- Foreseeable use not addressed in standard (also consider "reasonable misuse");
- Market or customer expectations (see Product Liability and General Product Safety in Chapter 2); or
- Specifics in directives (ERs) may exceed standards requirements.

European standards are the centerpiece of the total harmonization process (Figure 3-2). At the beginning of this chapter we mentioned the importance of using the proper European standards during the design phase in anticipation of the final conformity assessment. The first attempt at compliance (usually assessment of an existing product) will be a challenge, but after the first success the designer will be surprised at the amount of knowledge and understanding gained. The safety and EMC concepts learned can be utilized during future design phases, thereby ensuring successful testing and assessment to garner CE marking to all new products.

Compliance with European standards is defined as the *minimum* acceptable criteria and the manufacturer may have to go *beyond* the standards, such as complying with the EU laws (directives) and meeting market expectations.

The General Product Safety Directive 92/59/EEC

- "The purpose of the provisions of this directive is to ensure that products placed on the market are safe."
- "Where there are no specific Community provisions governing the safety of the products in question, a product shall be deemed safe when it conforms to the specific rules of national law of the member state in whose territory the product is in circulation."
- "In the absence of specific rules as referred to in paragraph 1 (above), the conformity of a product to the general safety requirement shall be assessed having regard to voluntary national standards giving effect to a European standard or, failing these, to standards drawn up in the member state in which the product is in circulation."

The Product Liability Directive 85/374/EEC

- "Protection of the consumer requires that all producers involved in the production process should be made liable, insofar as their finished product, component part, or any raw material supplied by them was defective."
- "The producer shall not be liable as a result of this directive if he proves that: . . . (d) the defect is due to compliance of the product with mandatory regulations issued by the public authorities."
- "To protect the physical well-being and property of the consumer, the defectiveness of the product should be determined by reference not to its fitness for use but to the lack of the safety which the public at large is entitled to expect . . . [e.g., harmonized or national standards]."

The Conformity Assessment Procedures and Rules for CE Marking Directive 93/465/EEC

- "The directives must set the range of choices which can be considered by the Council to give public authorities the high level of safety they seek, for a given product or product sector."
- "These procedures may only depart from the modules when the specific circumstances of a particular sector or directive so warrant. Such departures from the modules must be limited and must be explicitly justified in the relevant directive."

FIGURE 3-2 European Commission View on Standards for Safety Conformity

FIGURE 3-2 *(continued)*

The Global Approach to Certification and Testing (Commission Communication 89/C267/03)

- "The Council resolution of 7 May 1985 showed the way by accepting that there could be more than one means of proof of conformity to a directive. It provided for presumption of conformity to a directive on the basis of a European harmonized standard, or, during a transition period, of national standards which had been submitted and recognized as equivalent under Community control procedures."

Guide to the Implementation of Community Harmonization Directives

- "The concept of harmonized standards plays a important role in the framework of the New Approach Directives, in which it has a specific significance."
- "Conformity with national standards that have transposed harmonized standards, whose references have been published by the Commission in the *Official Journal of the European Communities,* confers a presumption of conformity with the essential requirements of the New Approach Directives."
- "Manufacturers are responsible for ensuring that the products they place on the market meet all relevant regulations [harmonized standards, etc.]."
- "If a product bearing the CE marking is found not to conform with the essential requirements laid down by the directives or if the standards were not correctly applied, action must be taken against the manufacturer."

Guidelines on the Application of Council Directive 73/23/EEC (LVD)

- Products are presumed to conform to the safety objectives of the "Low Voltage" Directive where the equipment has been manufactured in accordance with technical standards which, in the order laid down by the directive, are as follows:
  —European standards [EN or HD], which are referred to as harmonized standards in the Directive, drawn up in accordance with Article 5 by the bodies notified by the member states (in fact, these are standards made by CENELEC);
  —Where standards as defined in Article 5 have not yet been drawn up and published, the international rules issued by two international bodies, the International Commission on the rules for the approval of electrical equipment (CEE) or the International Electrotechnical Commission (IEC) (Article 6(1)), and published in accordance with the procedure laid down in Article 6(2) and (3); and
  —Where standards as defined in Article 5 or international standards as defined in Article 6 do not yet exist, the national standards of the member state, of manufacturer (Article 7).
- The standards referred to in Articles 5, 6, and 7, the application of which remains voluntary, provide a presumption of conformity for equipment manufactured in accordance with these standards.
- Alternatively, the manufacturer may construct the product in conformity with essential requirements (safety objectives) of the directive, without applying harmonized, international, or national standards. In such a case the product will not benefit from presumption of conformity conferred by the use of such standards and the manufacturer must include in the technical documentation (see Chapter 5) a description of the solutions adopted to satisfy the safety aspects of the directive.

(continued)

FIGURE 3-2 *(continued)*

Guide to the Application of 89/336/EEC Relating to Electromagnetic Compatibility (EMC)
• "European Harmonized Standards play a key role, not just because they significantly simplify the conformity assessment procedures (Article 10.1 of the directive) if used in full, but also because they provide by consent, a unique, harmonized, technical solution that has been based on an EMC analysis. This means that, even if those standards are not used (they are voluntary) in the design and manufacture of the apparatus, manufacturers should take them into account when performing their analysis."
• "The EMC directive is a new approach directive laying down apparatus protection requirements and leaving it to standards, primarily harmonized standards, to define product characteristics."
• "By way of information, Annex 7 contains a list of harmonized European standards that have been published in the EC's *Official Journal*. The application of the appropriate harmonized standards to an apparatus confers on that apparatus a presumption of conformity with the protection requirements of the directive. In other words, in the case of challenge, the responsible national authorities will have to prove that the product is not in conformity with the protection requirements. The presumption of conformity is conferred, in regulatory terms, only by the use of the national standards transposing a harmonized standard."

CHAPTER 4

Notified Bodies and Certification

Always do right. This will gratify some people and astonish the rest.

MARK TWAIN

Notified Bodies and Third Parties

Common European laws and standards are a good start toward harmonization, but mutual recognition of tests and certificates is necessary to create the requisite trust for total harmonization. Mutual recognition of test reports and certificates of one member state by another is the overriding requirement. A test is considered equivalent if performed by an accredited testing body on the basis of uniform assessment criteria. Accredited bodies are listed in the *OJEC* as *notified bodies* (safety), *competent bodies* (EMC), or referred to in general terms as *third parties*. As stated in the *OJEC,* the testing procedures may be undertaken by "specialist third parties, i.e., testing laboratories and certification bodies. These laboratories and bodies can, in turn be evaluated as to their technical competence by a third party (accreditation body) and hence be accredited . . . a company may entrust systematic or sample product testing to an independent body (third party)" (*OJEC*; 89/C 267/11). Therefore, the terms *third party* and *accredited body* (notified/competent) have the same meaning in Europe and should not be confused with nonaccredited parties, such as consultants, outside laboratories, U.S. agencies, manufacturers, and others that have not undergone an official evaluation by an EU accreditor and listed in the *OJEC*.

The goal of the accredited body is to protect the health of the consumer and the safety of the environment, with special emphasis placed on the well-being of the consumer (Figure 4-1). Placing consumer protection first also limits the product manufacturer's risks. When called upon, the testing and certification bodies must accurately assess equipment according to established European norms as the minimum criteria, confirming with such by the issuance of test reports and certifications recognized by all concerned parties. Doing anything less may put both the consumer *and* manufacturer at risk.

Role of the Notified Bodies
• Ensuring conformity, building consumer confidence, and protecting public interests;   • Provide facilities for conformity assessment on the conditions set out in the directives;   • Issue test reports and certificates on conformity;   • Provide guidance on the essential requirements or other provisions of the directives;   • Provide technical interpretation and applicability of standards;   • May assess manufacturer's technical file and documentation;   • Determine alternatives when no standards exist or other criteria used for conformity to ERs; and   • Test sample products for conformity and award "Approval Marks," with ongoing surveillance of production, to visibly demonstrate product quality and increase marketing advantage.
Guidelines for Notified Bodies
• Shall meet legally binding criteria set out in the annexes to the directives;   • Notified bodies are listed in the *Official Journal of the European Communities;*   • Member states should accept test reports and/or certificates of all notified bodies;   • Testing may be subcontracted to other laboratories, under the direct supervision of the body; and   • Notified bodies should reside in one of the EU member states.

FIGURE 4-1   Notified Body Rules

The European Commission has proposed a set of conditions to facilitate mutual recognition:

- Use common European standards (ENs) to replace national standards.
- Manufacturers can establish as much as possible quality systems that are certified by accredited bodies.
- Testing and certification bodies can be accredited to EN 45000 standards so that manufacturers can rely on independent and qualified testing, reports, and certifications.
- Certification bodies of the various countries agree on standards, thus permitting mutual recognition of tests and certificates.

Accreditation at the European level is a relatively new concept and dates back to the early 1980s. Accreditation means the recognition of the competence of a laboratory, certification, or inspection body by independent accreditors. The accreditation bodies are typically sanctioned by a member state government to perform audits of testing and certification institutions. The audited body must be able to demonstrate that it meets the criteria described in the annexes to the directives to become recognized at the state level and *notified* to the European Commission and the member states. The bodies that are notified are designated to carry out conformity assessment as set out in the directives. As an assessment technique, accreditation is an important tool to generate and maintain confidence in these bodies, just as certification is for products.

Accreditors in Germany, for both the regulated (mandatory certification) and nonregulated (voluntary certification) sectors, are united under the German Accreditation Council (DAR). In the nonregulated sector, the Sponsoring Association for Accreditation (TGA) is an association of German industry, commerce, insurance sectors, and other associations. The TGA is responsible for the inspection of testing laboratories and certification bodies. The TGA is joined by the club of European accreditors, which includes NAMAS (Great Britain), RNE (France), and others.

📖 Originally conceived as a government tool for consumer protection, the idea of accredited *notified bodies* and *product certifications* has received fresh impetus from the European Commission and many trade organizations.

Notification is the formal recognition of a testing and/or certification body by the European Commission. After a successful laboratory assessment and recognition by an accreditor, the state government notifies the Commission of the nationally accredited body. The accredited bodies are then listed in the *OJEC* as *Bodies Notified to the Commission and to the Member States,* commonly referred to as *notified bodies* or *third parties.*

Testing and certification bodies (third parties) must be qualified to an equal standard (EN 45000 and other criteria) if mutual recognition agreements (MRA) are to function (e.g., recognition of test reports, certificates, approvals). The aim is to ensure that accredited bodies operate in a comparable manner to build a solid foundation for confidence. The goal of accreditation is to ensure that recognized test methods are followed with verification achieved through mutual audits. The procedure ensures consistency and, hence, promotes greater competence and quality in testing and certification in the Community. This *network of national networks* is the most efficient means of achieving trust among all parties.

📖 The notified bodies' primary role is to provide conformity assessment and product certification for producers in a premarket capacity.

Market surveillance of products whether at the marketing and distribution stage or at the marketplace is the responsibility of the public authorities. In some member states, the notified bodies and the surveillance authorities come under the same authority, and lines of responsibility are organized to ensure that the two activities are separate and independent. Market surveillance may be performed at the national, regional, or local level, including the customs authorities. When a product's conformity is in doubt the authorities may resort to testing the product themselves at either their own testing facility or at those of a notified body. If they are testing at a notified body's facility, the final judgment is determined by the surveillance authorities, not by the notified body.

Testing, Certification, and Approval Marks

With respect to testing laboratories, certification, and inspection bodies:

There is an important role in ensuring conformity, building confidence, and protecting public interests. It has been estimated that there are over 10,000 testing laboratories and 1,000 certification bodies in Europe of varying capacity, legal status, and reputation. . . . Testing, certification and inspection may diminish the risks and, hence, the likelihood of damages (which in turn reduces the insurance cost), but do not affect the liability of the manufacturer. . . . Testing, certification, and inspection activities place the emphasis on preventing as far as possible the putting on the market of unsafe products, thus, avoiding damage being caused. (*OJEC*; 89/C 267/03)

European notified bodies (safety) and competent bodies (EMC) are accredited at the national level by the member states, such as in Germany; at the European level accreditation occurs when *notified to the Commission* and listed in the *Official Journal of the European Communities*. These accredited bodies are sanctioned by the European Commission and the member states to interpret directives and standards, and issue test reports and certificates on conformity. When a product becomes suspect or an incident occurs, the national enforcement authority may consider a test report or certificate issued by a European body. Having the notified body mark, certificate, and test report usually shifts the onus of proof in the manufacturer's favor, since the product was evaluated and certified by European recognized experts.

📖 A *certificate* is a visible form of attestation through an independent and impartial third party. Certification means that a product or service conforms to certain technical rules contained in standards or regulations.

An approval mark and certificate visibly demonstrate *product quality* (safety/EMC) and can lay to rest any doubts the consumer may have concerning a product's conformity to the EU directives and standards. The requirement for certification can come from:

- Consumers and users who often look for and expect a recognizable approval mark.
- Distributors who recognize the importance of certification for marketing and to reduce of risks.
- Product manufacturers who require the assurance of accurate testing for components they specify.

- Insurance companies who wish to assess and limit their risks.
- Government departments who wish to ascertain that a product complies with specific regulations.
- Various institutions and service providers, such as banks, schools, or medical institutions.
- Companies whose safety policy dictates a third-party attestation to reduce risks and insurance costs.

💣☀ It is often the distributor, importer, or authorized representative who is first called to task if an incident occurs or a product's conformity becomes suspect!

The *Commission's Guide to the Implementation of Community Harmonization Directives Based on the New Approach and Global Approach* states:

> Manufacturers are responsible for ensuring that the products they place on the market meet all the relevant regulations. Where these regulations do not require mandatory certification, manufacturers often seek *voluntary certification* to assure themselves that their products do meet the requirements set by law.

Defense of Due Diligence

Due diligence means taking all reasonable steps to ensure conformity. With only a few exceptions, the manufacturer/importer is ultimately responsible for a product's compliance to directives, but with a mark, certificate, or test report from a notified body the suppliers' risks are decreased and their defense of due diligence enhanced, should the products safety/EMC conformity come into question. It is the responsibility of notified bodies to accurately interpret directives and standards through testing and certification, thereby ensuring conformity and building consumer confidence.

☞ *Due diligence is a matter of record!* Over 100,000 products have received the "GS Mark" (GS = Safety Tested). Numerous Type-Approval Marks for components and EMC Marks have also been issued.

The increased marketing potential of certification and approval marks should not be overlooked. As the *Official Journal* states:

> Testing, certification, and inspection . . . can also be an integral part of a national industrial policy intended to promote goods both nationally and internationally. The reasoning behind such policies is that the reputation of certain certification marks [i.e., VDE/TUV/BG] represents a strong commercial advantage in international trade. (*OJEC*; 89/C 267/03)

The objective is to manufacture and sell safe products and not waste time arguing with customers who challenge a product's conformity. Designing and certifying products to meet the EU regulations may be more profitable than the oftentimes endless discussions and meetings within a company about the possible benefits. In the end, it's the customer's safety concerns that will determine whether the product is a success. Also remember that when a product becomes suspect, enforcement authorities will most likely refer the product to a notified body for testing and assessment.

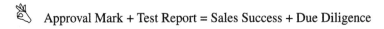

Approval Mark + Test Report = Sales Success + Due Diligence

The approval mark on the product and accompanying test report from a notified body serve two purposes. In regards to safety/EMC conformity, the test report is the manufacturer's best line of defense if the product's safety is questioned, and a reputable approval mark on the product is marketing's best sales tool! As the European Commissions *Guidelines on the Application of Council Directive 73/23/EEC (LVD),* states:

> In the case of challenge by the authorities in charge of market surveillance, a report in the sense of Article 8(2) (which, however, is not mandatory) is considered an element of proof. In fact, in addition to the three basic conformity assessment measures [technical documentation, declaration of conformity, and CE marking], Article 8(2) provides, in the event that conformity is challenged, for the possible submission to the market surveillance authority of a report drawn up by a notified body as evidence that the electrical equipment complies with the safety objectives. (Article 2 and Annex 1)

> The manufacturer or his authorized representative established in the Community may wish in certain cases to ask in advance for a report to be drawn up by a notified body in accordance with the procedure provided for in Article 11 and keep it together with the technical documentation. The availability of such a report would make matters easier and speedier in the event of challenge by the authorities.

When customers see an approval mark they know and trust (i.e., VDE/TUV/BG) they can be assured that the testing was performed to exacting quality standards and that an accurate test report and certificate exist. Figure 4-2 shows the hierarchy of European conformity acceptance. With today's liberalized approach (CE = self test), where testing and approvals for most products are no longer compulsory, the quality of the approval mark is more important than ever. There is a greater potential for abuse under the new system that must be counteracted by effective surveillance checks by the authorities, users, and competitors. Product manufacturers and suppliers should utilize testing and approval bodies with the

Type	Description	Advantages/Disadvantages
Approval Marks	The highest level for conformity verification and acceptance in Europe are the voluntary product safety and EMC "Approval Marks," issued by European notified and competent bodies. The "Approval Mark" is backed by an official test report and certificate from a European accredited body *(OJEC)*. *Note:* In Europe, "notified and competent bodies" have higher status than CB/NCBs (see CB Scheme below).	Advantages: • "Approval Mark" on product is visible to customers for increased marketing and sales potential. • Approval is viewed as a "Quality Marking" by consumers. • "Mark" verifies compliance of all products in series. • Tests per "EN standards"[1] (and IEC standards). • Official certificate and test report (TRF). • Technical file/documentation complete and accurate. • May support manufacturer's CE marking. • National enforcement authorities may consider marks, with report, as defense of "due diligence."
Official "Test Report"	Same as above except no certificate or mark	Advantages: • Official test report (TRF). • Tests per "EN standards"[2] (and IEC standards). • Authorities may consider report. • No factory inspection fee. Disadvantages: • Report limited to product/s tested (per S/N). • No "Approval Mark or Certificate" for potential customers and national enforcement authorities. • May only address specific standard(s).
"CB Certificate"	The "CB Scheme" is a mutual recognition of test results scheme for safety of electrical products. The CB Scheme is intended to "facilitate certification or approval" at a national level (see "Approval Marks" above). *Note:* To ensure compliance with all ENs and ERs for the "CE marking" requirements and expedite "Approval Marks," product manufacturers should use a national certification body (NCB) that is also a European notified body (in *OJEC*). Additional test samples and subsequent testing may also be reduced or eliminated for "Approval" submittal.	Advantages: • EU notified bodies may accept CB Certificate for issuance of notified body "Approval Mark" (above). • Some non-European NCBs participate in CB Scheme for their national approval. • Test report in CB format (TRF). Disadvantages: • No visible "Approval Mark" on product with only CB Cert. • Additional test samples may be required for national "Approval" submittal. • CB Certificates shall not be used in any form of advertising or sales promotion. • References "IEC standards"[3] and may not address EN standards for the CE marking. • May not address all ERs of the directives. • Limited to product tested (by S/N).

[1] European harmonized standards ("EN" listed in the *Official Journal*) shall be used for "CE" conformity verification

[2] In all cases the CE marking, declaration of conformity, and technical file are required.

[3] The "CE marking" is not a mark and should not be confused with a mark, certificate, or approval issued by a European recognized third-party (e.g., notified or competent body). The CE marking is a 'symbol' of the manufacturer's declaration.

© Lohbeck

FIGURE 4-2 Hierarchy of European Conformity Acceptance

FIGURE 4-2 *(continued)*

Type	Description	Advantages/Disadvantages
C€	"CE marking" is the manufacturers self-declaration symbol indicating conformity with the "essential requirements" of all relevant directives. The product CE marking and declaration of conformity is mandatory for most products. A technical file describing the product, design, assessment, tests, standards applied, rationale, etc., shall also be readily available.	Advantages: • "CE" indicates conformity to customs inspectors, allowing products to be "placed on the market." • Ensures "free movement of goods" within Community. • A "presumption of conformity" exists when products are in conformity with "EN standards."[4] Disadvantages: • No visible "Approval Mark or Certification." • Implies conformity with "minimum requirements" of the Directives and "CE" is not viewed as a quality marking. • Not for sales, marketing, or promotional purposes. • Authorities and customers may question CE marking based on incident, complaint, other.

[4] Terms; CE = European Conformity, EN = European Norm (EU harmonized standard), ER = Essential Requirements (of the Directives), CB = Certification Body, NCB = National Certification Body, S/N = Serial Number (tested sample[s]), TRF = Test Report Format, *OJEC = Official Journal of the European Communities.*

highest creditability and acceptance by both customers and authorities. Doing anything less may be a waste of money and time. Just because a notified body appears in the *Official Journal* does not mean the customer must accept it. Even the European Commission recognizes that "there are over 10,000 testing laboratories and 1,000 certification bodies in Europe of varying capacity, legal status, and reputation. . . ."

✓ The credibility and reputation of the bodies issuing "Approval Marks" are more important than ever before!

Figures 4-3, 4-4, and 4-5 illustrate examples of approval mark certificates issued by European bodies.

Fachausschuß Verwaltung
Prüf- und Zertifizierungsstelle
im BG-PRÜFZERT

Hauptverband der gewerblichen
Berufsgenossenschaften

Translation

GS Test Certificate

987123

no. of certificate

Name and address of the holder of the certificate: (customer)

COMPUTER COMPANY
ASICS WALK, SILICON VALLEY

Name and address of the manufacturer:

COMPUTER COMPANY
ASICS WALK, SILICON VALLEY

Ref. of customer:	Ref. of Test and Certification Body:	Date of Issue:
C-1234	987123	09.03.1998

Product designation: Personal Computer — - Visual display unit: MONITOR ABC
- Keyboard : KEYB 123
- System unit : CPU 333

Type: Tower ABCD-1234

Intended purpose:

Testing based on: Principles for occupational safety testing of electronic data processing equipment and visual display units (release: 01.97)
- IEC 950:1991 + A1:1992 + A2:1993 + A3:1995 + A4:1996,modified ;
  EN 60950:1992 + A1:1993 + A2:1993 + A3:1995 + A4:1997 (Safety of IT equipment)
- ISO 9241-3:1992; EN 29241-3:1993 (Visual display requirements)
- ISO 9241-8:1997; EN ISO 9241-8:1997 (Requirements for displayed colours)

Remarks: Graphic controller: XYZ-2000

The type tested meets the requirements specified in article 3 para. 1 of the Equipment Safety Act (GSG, § 3, Abs. 1). Thus, the type tested also complies with the provisions laid down in the directive 73/23/EEC (Electrical Equipment), amended by the directive 93/68/EEC.
The holder of the certificate is entitled to affix the GS-mark shown overleaf to the products complying with the type tested. At that, the holder of the certificate shall observe the conditions specified overleaf.
The present certificate including the right to affix the GS-mark will become invalid at the latest on:

31.03.2003

Further provisions concerning the validity, the extension of the validity and other conditions are laid down in the Rules of Procedure for Testing and Certification of October 1997.

Signature   Signature

Postal address:	Office:	Phone: 0 40/51 46-27 75
Deelbögenkamp 4	Deelbögenkamp 4	Fax: 0 40/51 46-20 14
22297 Hamburg	22297 Hamburg	

P2808e
10.97

In any case, the German original shall prevail.

FIGURE 4-3 Product Safety Certificate

Ⓝ **Nemko**

CERTIFICATE

No. P97555554

Order No. [illegible]

Applicant	Taiwan Export XX Taipei/Taiwan
Manufacturer	China Electronics ABC Taipei/Taiwan
Factory	China Electronics ABC Taipei/Taiwan
Group 63 92000	Build-in power supplies
Model/type	12345XXX
Data	5/2.5A 100-127/200-240V 50-60Hz
Other specification	Cl. I , BRAND NAME: SMITH; DC O/P:1-18A +5V, 0-12A +3.3V, 0-3A +12V, 0-0.3A -12V, 0- 0.3A -5V, 0-1.2A +5VFP, where, +5V & +3.3V output 105W max. and total power 145W max. Dots in model name can be any alphanumeric character or blank and denotes minor changes in SELV circuits or minor mechanical changes not affecting safety.
The above product is certified according to the flwg. standard(s)	Safety std.: EN 60 950 (1992) A1/A2/A3./A4, including possible listed national conditions/deviations for Norway.
Validity	The certificate is valid until 1 November 2007, provided that all signed certification conditions are complied with, and that possible changes to the product are notified to Nemko for acceptance prior to implementation. The validity time may be reduced in case new standards are made applicable. The certificate also applies as licence for use of Nemko's name and certification mark.
Remarks	The above certified equipment complies with current technical standard(s) in Norway regarding safety, as far as this can be checked. Compliance with requirements regarding building-in, protection against electric shock and Electromagnetic Compatibility (EMC) must be checked when the equipment is built-in a completed product or forms a part of a complete system. It should be noted that by marketing of equipment subject to official registration in Norway, the party responsible for the marketing shall be registered by the Norwegian Directorate for Product- and Electrical Safety.

Date of issue 28October 1997

signature
Geir Harthe

signature
Morten Smith

Nemko AS P.O. Box 73, Blindern N-0314 Oslo, Norway	Office address Gaustadalléen 30 Oslo	Telephone +47 22 96 03 30 Enterprise number:	Fax +47 22 96 05 50 NO 974404532

FIGURE 4-4   Component Safety Certificate

VDE Prüf- und Zertifizierungsinstitut
Zeichengenehmigung

Ausweis-Nr. Licence No.	Blatt Sheet
000000 F	2

Name und Sitz des Zeichengenehmigungs-Inhabers/*Name and registered seat of the marks licence holder*

Company XYZ

Fertigungsstätte/*Place of manufacture*

–AA– Factory ABC

Statistik/*Statistics*	Aktenzeichen/*File*	Datum/*Date*
1	12345–3251–0001/12345 TF1/	Datum

VDE–EMV–Zeichen / *VDE–EMC Mark* :

Die Bedingungen zum Benutzen des Zeichens sind auf Blatt 1 von < Datum > genannt.
The conditions to use the mark are referred to on page 1 of < date >.

Jahresgebühren–Einheiten/
annual fee units

Auf Grund der Prüfung nach den unten angegebenen harmonisierten europäischen Normen in Verbindung mit dem Artikel 10.1 der EG–Richtlinie 89/336/EWG vom 03. Mai 1989, umgesetzt in das EMV–Gesetz vom 09. November 1992 (Gesetz über die elektromagnetische Verträglichkeit (EMVG)), entsprechen die in diesem Ausweis aufgeführten Geräte den grundlegenden Schutzanforderungen zur elektromagnetischen Verträglichkeit (EMV):

Based on the test according to the harmonized European Standards as mentioned below, in conjunction with article 10 (1) of EC–Directive 89/336/EEC of 3 May 1989, transfered into German national law (Gesetz über die elektromagnetische Verträglichkeit (EMVG) vom 9. November 1992), the equipment described in this marks licence document meets the essential requirements of the electromagnetic compatibility (EMC).

Europäische Norm *European Norm*	Deutsche Norm *German norm*	VDE–Klassifikation *VDE classification*
EN 55022:1994	DIN EN 55022:1995–05	VDE 0878 Teil 22:1995–05
Grenzwerte für Einrichtungen der Klasse B / *Limits for class B equipment*		
EN 55022/A1:1995	DIN EN 55022/A1:1995–12	VDE 0878 Teil 22/A1:1995–12
EN 61000–3–2:1995	DIN EN 61000–3–2:1996–03	VDE 0838 Teil 2:1996–03
EN 61000–3–3:1995	DIN EN 61000–3–3:1996–03	VDE 0838 Teil 3:1996–03
EN 50082–1:1992	DIN EN 50082–1:1993–03	VDE 0839 Teil 82–1:1993–03

Produktbeschreibung: *Product description:*	Notebook Personal Computer	8,0
Typbezeichnung: *Type reference:*	PC 123	0,8
Nennspannung: *Rated voltage:*	DC 18 V (über externes Netzgerät / *via external power supply*) AC 100–240 V, 50–60 Hz	
Stromaufnahme: *Rated current:*	max. 1,5 A (bei DC 18 V / *at DC 18 V*)	
Schutzklasse: *Protection class:*	I / III	

– Fortsetzung siehe Blatt 3 / *continued on page 3* –

8,8

VDE Testing and Certification Institute · Institut VDE d'Essais et de Certification

VDE Verband Deutscher Elektrotechniker e.V.
VDE Association of German Electrical Engineers
Merianstraße 28 · D–63069 Offenbach

Telefon national (069) 8306-0 · Telefax (069) 8306-555
Telefon international (+49) (69) 8306-0

VDE-Form 77L.0596

FIGURE 4-5 Electromagnetic Compatibility Certificate

CHAPTER 5

Quality and Enforcement

Advice is what we ask for when we already know the answer but wish we didn't.

ERICA JONG

Product Quality versus Factory Quality

As recognized by the Commission, product quality is not automatically assumed (with or without CE marking) and must be earned:

> Besides using European standardization as a means of improving the quality and acceptability of products, it is necessary to enhance confidence in the ability of the manufacturer to supply quality products. This confidence cannot be simply imposed either upon public authorities or consumers; it depends primarily upon the public attitude of the manufacturer himself [i.e., it must be earned!]. (*OJEC*; 89/C 267/16)

The European and international standardization bodies have drawn up appropriate instruments (procedures, standards) to assist manufacturers and inspection bodies obtain consistent factory and product quality through proper management of their quality systems. EN 29000 (ISO 9000) describes techniques for manufacturer's quality assurance systems and factories, and accredited ISO 9000 auditors confirm these procedures through factory inspections and certification. The accreditation, competence, and quality of third-party testing and certification laboratories are promoted through the EN 45000 standards with their recognition granted by the various EU laboratory accreditors (i.e., DAR in Germany). The product's quality is shown by the various approval *certificates and marks* issued by these EU-accredited bodies (see Notified Bodies in Chapter 4).

Within the European Conformity safety/EMC context, three types of quality assessments are addressed in terms of the entities they relate to:

1. *Factory quality.* The manufacturer's system
2. *Laboratory quality.* The third parties testing competence
3. *Product quality.* The product's safety/EMC conformity

Certification of product quality by third parties is not mandatory for the vast majority of factories and products, however, certification (with approval mark on the product) is advisable because of customer expectations. All three quality types (factory, laboratory, and product quality), although coexisting, are in reality distinct spheres of work with assessments performed by separate groups of experts. To ensure customer acceptance for factory and product quality, manufacturers require independent verification through various certification schemes. In the first type (factory quality), an accredited party will assess a manufacturers facilities and procedures according to the ISO 9000 series and award a *certificate* to the compliant company. But, the accredited party does not give a product approval or CE marking. With *product quality* it is the product itself that is subjected to conformity assessment and certified according to product safety/EMC standards by an accredited body, and *an approval mark* is affixed to the product.

! A factory's ISO 9000 certification does *not* typically encompass product quality with respect to safety or EMC.

! A product's CE marking generally has *no relevance* to either factory or product quality.

The CE marking merely gives the impression that the manufacturer understands the laws and standards, but does not guarantee conformance with harmonized standards. Self-declarations, such as the CE marking and other conformity claims, are difficult to prove without independent verifiable support. Sales and marketing claims may be questioned by potential customers, especially when no independent evidence of compliance exists. It is what is behind the claim that counts! Trusting the initials CE to carry the product to success may not be enough. In the end, it is the consumer/users or end-product manufacturers who will make the decision by exercising their right to demand evidence of conformity to assure themselves that the safety/EMC quality of the product has met stringent European requirements.

The CE marking is *not* intended for marketing or sales and is *not* a quality marking of any kind. Confidence in a product's safety/EMC quality must be earned and verified through a system of independent checks and balances.

An approval mark on the equipment is always advisable, for if there is no mark on the equipment, the buyer may check for conformity, thus, increasing risk. To check the quality and conformity of a component, product, or machine, the consumer or end-product manufacturer may demand a *test report* for review and verification purposes. The manufacturer's technical file is not required, as it contains confidential information and is reserved exclusively for the authorities. The safety and EMC test reports however, should be readily available to anyone. After receiving the test report, you should seek a safety expert's opinion on the test reports accuracy and content.

⬭ If the test report is not available, the customer may *doubt the presumption of conformity,* just as an enforcement authority or testing body would.

As stated in the *Global Approach to Conformity Assessment,* "the national practices of certain member states in some industrial sectors rely on the manufacturer to ensure conformity to mandatory safety requirements, while others require third party (notified body) intervention. The choice of mechanisms to be applied and the procedures for applying them vary between member states and from sector to sector, as does the use of voluntary or mandatory certification. This variation is due to the relative importance attached to either the manufacturer's reputation; to the strength or weakness of national testing, certification, and inspection infrastructures; to national political traditions in respect of the role of legislation; and to different national attitudes towards product liability." (*OJEC*; 89/C 267/12)

📖 If a product becomes suspect, enforcement authorities will most likely refer the product to a notified body for testing and assessment. The manufacturer's best line of defense is a "test report" and "approval mark" from a notified body.

Product safety/EMC claims of conformity, for sales and marketing, often raise more questions than they answer and may not be supportable. Unsubstantiated claims are in abundance! Watch out for the following statements:

- "Product designed to meet . . ."
- "In accordance with . . ."
- "Complies with standard IEC . . ."
- "ISO 9000 certified . . ."
- "Lab-evaluated to EN 45001 . . ."
- "Tested in association with . . ."
- "Laboratory-recognized by . . ."
- "Testing witnessed by . . ."
- "Tested by U.S. lab for CE . . ."
- "Submitted for approval . . ."
- "The test report is on file" (… but we can't let you see it.)
- "The CE Mark is all you need!" (trust me . . .)
- "CE-certified" (ask me no questions . . .)
- "CE-approved" (. . . I'll tell you no lies!)

Consultants: Benefits and Limitations

Consultants and outside laboratories may be helpful depending on their experience, knowledge, cost, and—most important—the advice they give. There are some very good consultants, but you should be aware of their benefits and limitations. If you, the manufacturer, decide only to self-declare a product's conformity for the CE marking, then using the services of a reputable and experienced consultant may be beneficial. Consultants may improve the product manufacturer's chances of conformity, but beware of a false sense of security (I call it "the feel-good fallacy"). Even with a consultant, the product manufacturer will be liable for all the work performed (by themselves or the consultant) because it is the manufacturer whose name is on the product and the one who signs the declaration of conformity.

✓   Liability *cannot* be reduced by, or shifted to, a consultant.

Consultants and outside testing labs are *not* third parties. As mentioned, the European terms *third party* and *notified body* are interchangeable and signify *Bodies Notified to the Commission,* as listed in the *OJEC* (see Notified Bodies and Certification in Chapter 4).

A consultant may be knowledgeable in one field, but carefully check out those who offer the so-called "one-stop-shop" for U.S. & EU—Safety & EMC. Many of these shops actually specialize in only one area (such as U.S. EMC/FCC) and have a much lower level of competence in other areas.

Marketing and misleading advertising have reached a new high since the introduction of the CE marking. Some laboratories boast of their "affiliations" or "partnerships" with EU-notified and competent bodies, displaying certifications that attest to their "indisputable" qualifications and imply that their work is "backed" by EU partners. Certain U.S. laboratories have gone as far as purchasing some of the smaller EU-notified bodies in hopes of gaining access to both European conformity work *and* the manufacturer's pocketbook! However, external laboratory testing is supported by the EU body *only if the body directs the lab's work and issues a certificate and/or test report for the products testing.* Also beware of consultants who confuse you by selling their affiliations on one hand and telling you don't really need the EU body certificate on the other. If you use a consultant, seek out one with expertise who provides sound, reasonable advise and confirms such by obtaining both a *certificate and mark* from a European notified or competent body. This is the *best* protection for the consumer and you!

💣   Selling affiliations over certifications increases risks to the manufacturers and consumers.

EU bodies may subcontract certain testing activities to private laboratories. The rules concerning subcontracting of work by notified bodies to outside laboratories are strictly limited to specific tasks and stipulate that the notified body itself must perform the assessment, appraisal, and certification activities for which it is notified. Testing may be subcontracted to audited laboratories by the notified body, but only by "order" of the EU body and under direct *supervision and procedures* set down by the body. Only the European body's validated test report, certificate, and approval mark for the product in question has the support of the EU-notified or competent body.

!   Some consultants and test labs improperly issue so-called certificates. In Europe this right is exclusively reserved for EU-accredited certification bodies.

Compliance with all relevant directives and standards will not guarantee a safe product or prohibit authorities from taking action against products that are found to be dangerous, but verifiable conformity to the minimum requirements (e.g., the harmonized standards) goes a long way. Some consultants suggest using the directives' "general" *essential requirements* over the "specific" design and testing requirements contained in the *standards*.

●⁕   *Buyer beware.* Equipment's Conformity in Question! Some manufacturer's focus on the directives' *essential requirements* as a way to get around the safety/EMC rules (*standards*). In these cases there is *no* "presumption of conformity."

Most laws and standards are based on common sense and experience. By trusting the experts and garnering knowledge from them, you yourself can learn to apply the standards properly. With experience the principles become clearer, leading to cultivation of an almost inborn sense of how to interpret and apply the relevant requirements to achieve conformity. The best advice I can give is to submit a sample product or prototype directly to a notified body for a design review or to a reputable consultant who, in the end, also obtains the safety/EMC certificate and mark. In many cases the evaluation can also be performed at the manufacturer's facility. In the end, it usually saves you considerable time and money if you use experts, especially when you factor in the time it may take you to understand the standards and document the assessment. Among the many benefits, the notified body does the complete assessment and tests, identifies product deviations, and generates a report in a fraction of the time it would take you to do it yourself. Also, you can be sure the assessment and documentation are accurate and accepted by both customers and authorities.

U.S. and EU Differences

United States (U.S.) and European Union (EU) standards and safety philosophy are not equal; interpretation and application of the European standards in the United States are oftentimes inconsistent with the EU experts' interpretation and application. The differences in safety philosophy are a matter of history, culture, and environment and frequently make it difficult for safety engineers operating on one continent to comprehend and accept the views of safety engineers on the other continent.

💣☀ These philosophical differences affect the testing results, even when the same standards are applied.

The requirements contained in the European standards and the level of testing by the European bodies are recognized as stricter (especially in Germany) than their U.S. counterparts. It is generally acknowledged that products meeting the European standards exceed U.S. standards, and a product that only meets the U.S. standards almost certainly does not comply with the EU requirements. This is generally true in most areas of product and component safety, machine safety, and EMC. In regard to safety, there are numerous technical differences between the U.S. and EU standards. Safety guidelines under conflict include: components; construction; spacings; guarding; fault and ground tests; labeling; manuals; documentation; and warnings. In regards to EMC, the test configuration and limits vary between the United States and European Union. Most notably, Europe requires additional immunity tests and more product categories covered.

Contrast between the EU and U.S. safety views are most notable, but not limited to, the followings areas:

- Shock versus fire;
- Construction versus testing;
- Office/industrial environment (clean/dust and water) versus outside/inside (rain/no-rain);
- Production: hi-pot and ground testing versus hi-pot or no-tests; and
- Liability: "approved" versus "listed."

The United States and European Union place a different emphasis on safety hazards. "Fire" hazards are the most important safety consideration in the United States and "shock" hazards are of the highest priority in Europe. Fire statistics between two comparable-size U.S. and EU cities show a interesting trend, with five to ten times more incidents of fires in U.S. cities. The underlying reason for this is the difference in voltage and current (e.g., the mains' power supplied to products). For example, the power in the United States is about 115 volts for portable products, with double the input current (i.e., 6 amps), whereas in Europe the voltage is double that of the United States, about 230 volts, with half the current (i.e., 3 amps). Hence, with the higher voltage in Europe there is an increased risk of *shock* (EU, 230 V, vs. U.S., 115 V). With double the current in the United States the risk is primarily *fire* (U.S., 6 A vs. EU, 3 A).

High voltage in Europe ➡ Shock hazard!

High current in United States ➡ Fire hazard!

$6\!\!\!\!\sigma\!\!\!\!\int$ The requirements contained in the "standards" and the safety experts' "interpretations" reflect wide differences of opinion between U.S. and EU views toward safety.

You must look at the history of UL's fire enclosures and plastics flammability rating system to understand the U.S. view. Historically, the North American standards and agencies (UL/CSA) focused on flammability, plastics, wire insulation, and enclosures. Emanating from this evolved the theory that during fault tests an internal fire is permitted within the product, just as long as it eventually goes out (fire shall not propagate beyond equipment) assuming the product is constructed of flame retardant materials and has a "fire enclosure." To exemplify this concept, the enclosure bottom openings in computers and similar products are restricted to a maximum diameter of 2 mm (0.08 in) to limit the spread of fire. With this concept it is assumed that dripping flaming plastic will self-extinguish prior to exiting through the bottom openings. If the openings exceed the size limits, then the product is subjected to UL's infamous *flaming oil* test whereupon flaming oil is poured through the interior of the product and onto an awaiting test cotton below. If the cotton ignites the product fails the test. Therein lies the fallacy: the cotton usually ignites, thus, forcing the product to be redesigned with small openings (facts are stranger than fiction)! The North American concept of allowing a limited burn within a product is unacceptable to many EU testing bodies and is, therefore, not allowed. Instead EU standards require a fuse or other protective device to activate to "prevent any fire" and not merely contain it as U.S. philosophy allows.

$\Box\!\!\!\!\!/$ European safety theory says "fire is not permitted!" U.S. standards impose fire enclosure requirements and "allow limited fire" during testing.

In stark contrast, the EU safety standards and experts pay special attention to "shock hazards" through separation of circuits, components, isolation, transformers, and PCBs to keep the high voltages isolated from the operator. Additional construction requirements and tests are also employed for short circuits (single faults), protection of the service personnel, guarding, and so on.

$\Box\!\!\!\Box$ Obtaining European standards is the easy part; understanding the requirements is another story! Misapplication and/or misinterpretation are commonplace.

Applying U.S. standards or safety methodology to product assessment increases the risks of nonconformity because there are fundamental technical and interpretation differences between the U.S. and EU views (ANSI/UL vs. EN/IEC). In the area of safety philosophy, the U.S. test agencies, known as Nationally Recognized Test Laboratory's (NRTL), stress *fire* and *tests,* whereas, the European bodies focus on *shock* and *construction.* Both camps generally consider all safety aspects during a product assessment, but because of the major philosophical and technical differences between the U.S. and EU there are oftentimes different test results or findings. Because of these fundamental differences in the concept of safety, products designed to meet only the U.S. standards rarely meet the European requirements.

!   Meeting the European standards and interpretations is the minimum acceptable criteria.

The U.S. and EU standards deal with environmental "dust and water" requirements, for industrial areas, from different perspectives. In the European standards, ingress protection (IP), via dust and water tests, is mandatory for most machinery and some products. The dust tests are especially important for shock protection since dust found in industrial environments often contains metallic particles. Dust with even small amounts of metal particles can cause shock hazards. This can occur when dust is deposited over high-voltage circuits within the equipment that eventually cause high voltage to arc over to touchable secondary circuits or the enclosure, presenting a danger to operators. When EU standards require ingress protection (i.e., IP54), i.e., for machines, both dust and water tests are performed (see Design Guide—Enclosures in Chapter 6). In the United States there is a tendency to focus on water testing (NEMA) alone or in place of dust testing. Even water testing may not be required in the United States unless the equipment is intended for outside use or use in extreme industrial environments. Performing a water test in place of the dust test is not unheard of in the United States but is not permitted in the EU standards.

📖   Substitution testing may not take the place of the required tests. Frequently, equipment passes water testing but fails the dust test.

Another area where U.S. and EU differences exist is in production testing. In addition to a product's initial safety assessment and tests (type tests), all electrical products and machines must undergo routine electrical safety testing during the production process (see Design Guide—Production Tests in Chapter 6). These tests are performed to detect any safety defects in the manufacturing process and materials. As a minimum, the European safety standards and certification bodies typically require two production tests for equipment (i.e., computers, machines) and components (i.e., power supplies), one for electrical strength (hi-pot), and another for resistance of protective earth (ground continuity). For example, European standards EN 60950 and 50116 (Type Tests and Routine Tests for Information Technology Equipment) require these two production tests. In the United States, however, only the hi-pot test is required for most NRT-listed products, even when these are certified to International standards (i.e., IEC 950/EN 60950). Furthermore, hi-pot and ground tests *may not be required* (no-tests) in production for some critical components, such as power supplies, even when they bear a NRTL-recognition mark.

U.S.-NRTL safety marks are restricted to *listings* for products and *recognitions* for components. *Listing* or *Recognition* means the product or component was evaluated and passed the tests, but they are not approvals. The European *approval* designation (certification), used in association with a NRTL listing or recognition, is often discouraged in the United States to limit any implied warranty. Furthermore, the wording of some important terminology in the U.S. versions of IEC standards have been changed to reflect U.S. legal considerations. The following passages illustrate differences in the U.S. and EU interpretations of the same "base standard," Europe's EN 60950/IEC 950 versus the United States' UL 1950 (Annex NAD):

Europe	United States
"avoid the risk"	"reduce the risk"
"remain safe"	"not introduce a hazard"
"prevent a hazard to"	"reduce the risk of injury of"
"protection against"	"reduction of risks of injury due to"
"ensure protection against"	"reduce the risk of"
"the operator shall be prevented from having access to:"	"means shall be provided to reduce the risk of the operator having access to:"
"to provide adequate protection against the risk of personal injury"	"to reduce the risk of injury to persons"
"so constructed that no dangerous concentration of these materials can exist and that no hazard within the meaning of this standard is created"	"so constructed to reduce the risk from dangerous concentration of these materials and to reduce the risk created"
"ensures that such testing will indicate that the assembled equipment conforms to the requirements of the standard"	"indicates that the results of such testing will be representative of the results of testing the assembled equipment"

The European bodies (especially in Germany) issue certified approvals and marks according to clearly defined standards, which are definitive statements on their assessment results. The European *approval* term means that the equipment bearing an approval mark has been successfully tested and passed, but more important, approvals are perceived by users and customers as a guarantee that the equipment is *safe for use* (see Notified Bodies and Certification in Chapter 4).

It is not just using the correct standard that counts; more important it is proper interpretation. Customers, users, and authorities have the right to question, "Who did the assessment, self-test or EU third party (notified body)?" and "Does the product really comply?" To answer these questions and allay customer's fears many companies have their products tested in the U.S. by an outside lab, NRTL, or consultant in accordance to an international standard (i.e., UL 1950 ≈ IEC 950). Some companies feel that this U.S. testing is a substitute for European conformity testing by EU third parties, but this could not be further from the truth. Other companies wisely take the sound approach and limit their risks by submitting products to notified bodies. The European labs have much experience assessing products that have already been tested and certified (listed or recognized) by U.S. labs to their so-called equivalent UL/IEC standard. The results have been surprising and disconcerting, with the vast majority of these pretested products failing the requirements when tested by a European accredited body (i.e., VDE/TUV).

!   It is this author's opinion that the majority of components, products, and machines fail safety tests when assessed by a European notified body, even when they were pretested to international standards (EN/IEC) by the manufacturer, consultant, or U.S. agency (see Enforcement below and CE's Credibility at Risk in Chapter 7).

Problems that are often encountered with U.S. safety testing according to European requirements include:

- Inappropriate standard applied;
- Applying one standard when several are applicable;
- Safety issues unsatisfied when not addressed in standard at hand;
- Not considering all requirements for environment (industrial, hazardous locations, moisture);
- Focus on *fire* and *tests* (U.S.) instead of *shock* and *construction* (EU) requirements;
- Risk assessment inadequate or not performed;
- Testing or fusing in lieu of construction and other requirements;
- Enclosure and hi-pot test (box-it and pot-it!) to cover nonapproved components;
- Production testing inadequate or not performed;
- Warning labels used where guarding, interlocks, or other fail-safe measures required;
- Manuals, warnings, labeling, and languages not in order;
- Member state national requirements (PTT, VBG, ZH, DIN, VDE) not addressed or referred to; or
- Essential Requirements of the relevant European directive(s) not considered.

To stay competitive in the world market and strive for the highest levels of safety, the European manufacturers and standards and testing bodies have been very willing to accept new ideas. In addition to Europe's strong shock protection philosophy, the UL flammability and enclosure requirements have now been incorporated into their standards, making the EN/IEC standards truly world-class safety requirements. The good news is that the United States is moving in the direction of EU standards and may eventually adopt them, but there's a long way to go. Unfortunately, it will take some time for North America to understand the European safety philosophy of shock protection and accept their interpretations, just as it has taken time for Europe to understand the U.S. flammability concepts. Change takes time, as the following witticism suggests:

☺   The United States is going to the metric system, but is getting there inch-by-inch!

The Quality and Safety Mindset

Exporters should be aware that the German market is the single largest market in Europe and amounts to approximately one third of the entire European Union. The German economy is rich, sophisticated, mature, and its people expect quality and safety in the products they buy and sell. They want to buy the best and are willing to pay for it. Selling products in Germany takes much more than a sales pitch. Other than Scandinavia, no other country has a more regulated and formalized system of showing conformity to the technical rules. These rules have not been erected to keep products out, but in the case of Germany are an inherent part of their historical situation and structure. The standards and rules they have set concerning design, safety, and reliability have served as a benchmarks in other markets. The German safety/EMC laws and standards have become models for the new European directives and harmonized standards. Furthermore, the German national laws and standards are now in line with Europe's (and vice versa).

Consumer expectations vary considerably within Europe. To illustrate this point, there was a saying that described the differing attitudes between the United Kingdom and Germany: "In Germany, everything is forbidden except that which is permitted. In England, everything is permitted except that which is forbidden."[1] This is obviously an overgeneralization, but nevertheless, the principles are accurate. The United Kingdom has a long tradition of self-certification, whereas, third-party certification has been quite common in Germany, even though not mandatory by law for most products. In Germany, certification for many products is voluntary, but the need for a product to certified by a reputable third party, one they know and recognize (VDE/TUV), is often a marketing requirement because of the consumers' greater awareness of standards. Voluntary certification is actively encouraged by the government, manufacturers, insurers, and consumer groups. They expect high-quality and safe products and look for third-party approvals on the products they buy.

Technical requirements and safety/EMC laws and standards exist to protect the consumer from dangerous occurrences caused by products. The German Equipment

---

[1] This section is based in part, with quotations or facsimiles contained herein, on the DTI-Department of Trade and Industry (UK) brochure; "Exporters Guide to Technical Requirements in the Federal Republic of Germany."

Safety Law (GSG) of 1968 says that "all products must be safe" and meet the "recognized rules of technology." A product's conformity with the relevant standards is "deemed to satisfy" the law ("presumption of conformity"). German national laws such as the Accident Insurance Act are in place to encourage product manufacturers to supply safe products and enforce conformity through the users and local authorities. The Accident Insurance Act requires accident coverage to be taken by employers from labor organizations, known as the BG, that cover workers for industrial injuries. It is a legal requirement that both the employer and employee comply with the Accident Prevention Regulations. The German safety laws include provisions for enforcement at the state level. None of these laws or regulations contradict the European directives.

The German and European policies have raised the awareness of quality and safety for the buyer. The directive on the Global Approach recognizes this: "The reasoning behind such policies is that the reputation of certain certification marks represents a strong commercial advantage to commercial trade" (*OJEC*; 89/C 267/12). The so-called technical barriers to trade are not new and have developed over decades and have only recently become apparent with enforcement of the CE marking. The requirements were not erected to keep products out but more for a global harmonization to eliminate any barriers to trade and raise the level of all products "quality and safety."

Market Surveillance

Without effective market surveillance manufacturers begin to ask themselves why they should go through the costly process of CE marking when many of their competitors ignore it and get away with it. Effective market surveillance at national and member state levels is a prerequisite for consumer confidence in the Single Market. Surveillance and enforcement are still in their infancy in Europe and have a long way to go. The results of market surveillance have been spotty at best, with some notable exceptions in countries such as Finland, Germany, and Sweden. The European Commission publishes directives and standards that give member states the necessary tools for surveillance and enforcement, but implementation is left to the national authorities. Most important, uniform market surveillance is established to achieve a "high level of safety protection" for all citizens and build "consumer confidence" in the products they buy. Trust is good, but effective control is better! We must recognize that manufacturers sometimes make mistakes. Some EU countries only check for the CE marking and declaration of conformity "at the border," which does not ensure a product's conformity; more than 50% of products with CE marking fail to meet safety and EMC standards. If EU consumers cannot trust products from abroad they will continue to buy those manufactured locally. Safety should not stop "at the border" (1 and 2 below) with only documentation checks. Market surveillance will only work if enforced by local safety/EMC *experts* "on the ground" via product audits and testing (3 and 4 below).

The surveillance levels follow, with higher numbers indicating increased surveillance effectiveness:

Surveillance at the border by customs inspectors and authorities

1. *Administrative.* Check for CE marking and declaration, etc.
2. *Product.* Brief review of labels, accompanying documentation, etc.

Surveillance on the ground by authorities with support of safety/EMC experts

3. *Administrative.* Detailed review of declaration, CE marking, manuals, etc.
4. *Product.* Assess and test product per standard(s), etc.

All member states should establish "easily identifiable" and "effective surveillance" authorities and experts. A proactive approach to market surveillance should be promoted to limit the risks of dangers from nonconforming products. The objectives of market surveillance must be active and effective enforcement, rapid communication on defective products, and equal protection for all citizens. Just as "standards" are the foundation for safe product design, so is effective "market surveillance" the cornerstone for securing consumer confidence in the Single Market.

Enforcement

The Product Liability Directive (85/374/EEC) on victims' rights states that a product must meet a "high level of safety" that the consumer may reasonably expect. It is generally understood that harmonized standards are the "minimum safety criteria." However, a product's compliance with the harmonized standards does not always guarantee a safe product. The manufacturer may have to exceed these requirements to make a truly safe product—one that also meets the customer's expectations. Doing anything less may put the consumer at risk and increase the manufacturer's likelihood of damages.

"Strict liability" was established in the European Union to protect consumers (users and operators) from defective products. Directive 85/374/EEC changes the old approach of proving negligence to a new emphasis of strict liability on the manufacturer's part. Consumers can now initiate civil actions themselves, without the need to prove negligence. All producers involved in the production process are liable, insofar as the finished product, component, or raw material they supply is defective. The consumer can take simultaneous action against all parties involved in the supply chain. Moreover, the directive does not set any financial ceiling on the manufacturer's liability.

A questionable product may come to the attention of the enforcement authorities by way of a customs inspection, market surveillance audit, user complaint, competitor, or incident. In any case, the national enforcement authorities rely on either the harmonized standards or their national transposed standards to assess the suspect products conformity or lack thereof.

When a member state ascertains a product's nonconformity, it may take appropriate measures. Depending on the situation and nonconformity of the product, the manufacturer will then have to do one or more of the following with the product: restrict or prohibit its sale, make it comply, withdraw it from the market, or destroy it. A domino effect then occurs. The member state notifies the Commission, which, in turn, notifies the other member states, which then also take appropriate action. Figure 5-1 shows the risks of non-conformity. Three examples illustrating enforcement trends for EMC, product safety, and machinery are as follows.

Example 1: EMC Enforcement in Europe

The European EMC authorities have been very active since the EMC directive came into force. So far, the most rigorous enforcement has been in Germany and Sweden, with the United Kingdom, Italy, and others to follow suit. The authorities are checking manufacturers' technical documentation, declarations, and testing products, resulting in further investigation, prosecution, and (in some cases) product withdrawal from the market. In Sweden products are removed from the market as soon as they are identified as noncompliant. Estimates of noncompliant products range from 30 to 50%, depending on the country and industry sector (and whom you speak to). During 1996 in Germany alone, 1,500 technical documents were reviewed each month and 750 were passed to regional offices for further investigation. Many companies have received fines. Of the products tested 61% had incorrect declarations of conformity, 28% had problems relating to the CE marking, and 11% had technical shortcomings. Germany's version of the European EMC directive, known as the EMVG, allows enforcement authorities to impose fines for nonconformity without taking companies to court. This helps to fund further enforcement activities in Germany. The German market for products is about one third of the entire European market. These examples are only the tip of the iceberg, and numerous companies are rapidly upgrading their products before enforcers can conclude their investigations (ref; Approval [UK], Mar/Apr 1997, pp. 4–5). In addition, products sometimes meet the EMC limits with minimal margins (it's good enough!), and manufacturers start CE marking and shipping without keeping in mind that marginal products oftentimes fall out of compliance after they are put into production because of manufacturing differences and design changes. The EMC directive requires spot check testing of products after they are put into production to verify ongoing conformity, but this retesting requirement is sometimes forgotten.

☹ Products on the market for some time become easy targets for market surveillance authorities that realize that reverification testing for ongoing EMC conformity is only sporadically performed by some manufacturers.

- **Low-Voltage Directive (LVD)—73/23/EEC**   **CE Marking by 1/1/97**
  [electrical products 50-1000Vac, 75-1500Vdc]

  If a member state prohibits a product to be marketed for safety reasons, such as nonconformance with or *faulty application of standards,* or failure to comply with good engineering practice, the member state shall immediately inform the other member states. If CE marking has been affixed unduly, the manufacturer or importer shall be obliged to make the product comply. Where non-compliance continues, the member state shall restrict or prohibit its sale or ensure that *the product is withdrawn from the market.*

- **Electromagnetic Compatibility (EMC) Directive—89/336/EEC**   **CE Marking by 1/1/96**
  [electrical products, machinery, etc.]

  Where a member state ascertains that a product does not comply with the protection require-ments, it shall take all appropriate measures to *withdraw the product* from the market, prohibit its placing on the market, or restrict its free movement and shall immediately inform the Com-mission for reasons of noncompliance such as failure to satisfy requirements or the *incorrect application of the standards . . .* The Commission can then inform the member states and take the appropriate action against the author of the attestation (declaration).

- **Machinery Directive—89/392/EEC**   **CE Marking by 1/1/95**
  [includes mechanical and electrical safety]

  When a member state ascertains that machinery bearing the CE marking, is liable to endanger the safety of persons it shall take all appropriate measures to *withdraw the machine* from the market and prohibit its marketing. If, after consultation with the parties concerned, the commis-sion considers the measure is justified, it shall immediately inform the other member states and the member states shall take appropriate action against whomsoever has affixed the CE marking.

- **General Product Safety Directive—92/59/EEC**

  Allows only "safe products" to be placed on the market and meet a "high level of protection" of safety. Gives member states authority to control the safety of products and take the appropriate measures. Sets up a system of *rapid exchange of information on defective products* at national and Community levels. Information is passed to all member states and possible publishing in the *Official Journal* (negative PR).

- **Product Liability Directive—85/374/EEC**

  To better protect the consumer, "strict liability" (American-style law) now applies to producers, importers, or distributors of any material, component, product, machine, etc. A product is con-sidered "defective" if it does not provide the safety a person is entitled to expect. The consumer may initiate *actions against all parties* involved in the supply chain. The injured person has only to show a casual relationship between the defect and the damage, and the producer/importer may be assumed guilty.

- **CE Marking Amendment—93/68/EEC**

  Where a member state (or competent authority) establishes that the CE marking has been affixed unduly, the manufacturer or their agent established within the Community market shall be obliged to make the product/equipment comply as regards the provisions concerning the CE marking and to end the infringement under conditions imposed by the member state. Where noncompliance continues, the member state must *take all appropriate measures* to restrict or prohibit the placing on the market of the product/equipment in question or to *ensure it is with-drawn* from the market.

  © Lohbeck

FIGURE 5-1   The Risks of Non-Conformity

Example 2: Product Safety Enforcement in Finland

Product safety experts can usually spot noncompliant products visually in a matter of seconds or minutes without removing a product's cover. Minor discrepancies found on the outside of the equipment often mean that bigger problems are within. For example, over a two-year period the Finnish enforcement authority obtained over 1,200 products from stores and importers for market surveillance purposes. The inspectors selected products that appeared to have external deviations, such as wrong nameplate, incorrect plug or cord, user manual missing, or unknown trademark. The results of the audit testing were as follows:

- 24% had minor deviations from the standards (76% with "CE marking");
- 44% had deviations that may endanger safety (66% with "CE marking");
- 8% had serious deviations (57% with "CE marking"); and
- only 24% met the relevant safety standards (65% with "CE marking").

Over 75% of the 1,200 products tested failed to meet the requirements. Depending on the audit findings, the resulting actions taken by the Finnish authority varied from written warnings to product withdrawals or a total ban on sales. It is common knowledge among the EU-notified bodies and enforcement authorities that minor deviations seen externally on the equipment, such as wrong information on nameplate, incorrect indicator light colors, actuator symbols or warning labels not per standards, improper cord/plug/strain relief, user manual missing or not in proper language, usually means major deviations internally that could affect safety.

💣 Market surveillance principle: Minor *external* deviations, usually means major *internal* deviations!

Example 3: Machine Safety Enforcement in Sweden

As an example of machine safety enforcement, in 1996 the Swedish government carried out 2,000 workplace inspections. The focus was on machines installed after the CE marking deadline. The following items were questioned and checked: (1) Was a CE marking affixed? (2) Was a declaration of conformity in place? (3) Were the operator's instructions in Swedish? (4) Which notified body was involved for Annex IV machines? The results were much worse than expected with 50% of 3,000 inspected machines having one or more deviations (and they were only checking paperwork, this time). The employers received an inspection report and the machine suppliers received official letters describing what must be done to rectify the imperfections. In addition, 20,000 more companies received letters outlining the requirements. They were also informed that they could look forward to an inspection in the future. The objective of the campaign was to get the Machinery Directives provisions to be effective in practice. The Swedish government is actively lobbying the European Commission and other member state authorities to increase enforcement efforts throughout Europe.

✓ Over 50% of 3,000 machines inspected in Sweden failed to comply with documentation and labeling rules. The authorities were only performing administration checks for the CE marking, declaration of conformity, and manuals this time—more to come!

Legal Requirements and Market Demands

The European Commission's goal of "single-market access" is virtually a reality for companies who follow the New Approach rules. Technical requirements (standards) and laws (essential requirements) exist for the most part to protect the public from interference (EMC) and dangerous occurrences (safety) caused by products. The CE marking is a symbol of the manufacturer's self-declaration to indicate conformity with the minimum requirements, allows products to be "placed on the market," and "ensures the free movement of goods."

Keep in mind that the CE marking is not intended to be a sales or marketing tool, for "if all products have the CE marking, *what places one product above another?*"

In the past, enforcement was at best spotty, but now with the New Approach everyone is becoming aware of the requirements—manufacturers, buyers, competitors, and inspectors alike. The CE marking is primarily for market control by customs inspectors and enforcement authorities, but customers may expect more. The Commission has laid down the minimum requirements, and this leaves it up to the market forces to set their own expectations. In doing so the manufacturer has a choice of meeting the minimum legal requirements (ERs) or meeting the market expectations as well. In Europe expectations vary considerably between countries. Even with self-declaration, the demand for third-party certification and marks is unlikely to go away, especially in the more regulated countries like Germany and Scandinavia. Certification and marks may be a marketing requirement, like ISO 9000, because of the greater awareness of standards among consumers at all levels in the European Union. Successful product marketing requires an awareness and desire to meet the customer's expectations. The customer knows that buying products bearing reputable third-party *marks,* such as VDE or TUV approval marks, ensures that the product is not only in accordance with the European directives and standards, but also is "suitable" for its purpose and, above all, "safe."[2]

---

[2] *Note.* See Chapter 6 for product safety technical requirements, Chapter 4 for safety verification, and Chapter 7 for the definition of safety.

FIGURE 5-2 We need a plan!

Design Guide for Safety Conformity

It is not who is right, but what is right, that is important.

THOMAS HUXLEY

Introduction: Electrical Safety for Products and Machines

This chapter is a step-by-step approach for equipment safety design and assessment. It describes the principles of safety and requirements as found in the European harmonized standards. Also included, is an easy-to-follow safety checklist at the end of the section. The chapter is a basic design guide for electrical products and machinery that concentrates on component selection and construction techniques for product safety. The information is generally applicable to most product types such as information technology equipment (ITE), test and measurement devices, appliances, machinery, and other similar equipment. The design tips detailed concentrate on problematic areas that manufacturers most often encounter during their first safety assessment. This chapter will prove to be an invaluable aid to understanding the safety standards and identifying deficiencies in products. Its goal is to give equipment designers and manufacturers a better understanding of European and international safety considerations, including the unique European safety philosophy (see also U.S. and EU Differences in Chapter 5). Understanding this safety philosophy and adhering to a few simple rules will help product manufacturers to achieve safety conformity on all products manufactured for worldwide export.

Guide limitations. The appropriate product or machine safety standards take precedence over the guidance and design tips presented in this chapter. A complete assessment, according to all relevant standards, must be performed by qualified persons to ensure conformity.

The terms *product* and *equipment* are interchangeable and include machinery. *Operator, user,* and *consumer* are also equivalent terms. The term *appliance* is generally applied to household and similar products and includes commercial appliances. I must caution you that even when a product complies with these design tips, the product may not necessarily conform to all aspects of the standards.

Proper application of the guidelines in this chapter, along with the standards, will better ensure that the product complies with European and international safety standards (EN/IEC). The design guide in this chapter is for product, machine, and system designers. It does not cover testing or assessment of individual components (e.g., built-in components). Component standards exist to evaluate such components and should be referred to when necessary. This guide stresses the importance of selecting components that have visible European approval marks, such as TUV or VDE, as positive evidence of compliance, in lieu of testing each component, for which no evidence exists. Some component standards are found in the Low-Voltage Directive (LVD) and their conformity is mandated by the Low-Voltage and General Product Safety Directive. CE markings are not allowed for most components and, where found, are only self-declarations and are not evidence of compliance from an independent European body.

Low-Voltage Directive (LVD) of 1973 is the primary directive for safety compliance of electrical equipment. The Low-Voltage Directive was a real groundbreaker in European standardization and certification because of the numerous EN and IEC standards published setting the technical rules for equipment and components. The LVD covers safety of all electrical products that operate at 50 to 1,000 Vac or 75 to 1,500 Vdc. Products covered by the LVD include Information Technology Equipment (EN 60950), Test and Measurement Devices for Lab Use (EN 61010-1), Electronic Apparatus for Household Use (EN 60065), Household and Similar Appliances (EN 60335), and others. The appliance standards also cover "similar" products that are for residential or commercial use, such as deep fat fryers, cooking ranges, electric pumps and generators, clothes dryers and washers, floor cleaning equipment, and many others. See Equipment Classification for more examples.

Electrical safety is not limited only to products covered by the Low-Voltage Directive (LVD). According to the Machinery Directive, machine designers must also be aware of electrical safety since machines pose electrical hazards and they utilize numerous electrical components and subassemblies. The Low-Voltage Directive, General Product Safety Directive, and Machinery Directive mandate a product's conformity with the relevant electrical, mechanical, component, and other safety standards. To ensure that machinery electrical hazards are addressed by machine manufacturers, EN 60204-1/IEC 204-1 (Electrical Equipment of Machines) was published in both the Low-Voltage and Machinery directives, and, therefore, electrical requirements must also be applied to machinery. EN 60204-1 is a generic safety standard (type B) used in conjunction with the relevant machine safety standards (type C).

A machine that complies with EN 60204-1 (IEC 204-1) is presupposed to conform to the requirements of the Low-Voltage Directive.

Some people assume that the term *low voltage* means safe voltages, but as you have seen the LVD applies to products that operate at typical line (mains) voltages (e.g., 230/400 volts in Europe and 120/208 volts in the U.S.) that present shock and fire hazards. Safe voltages are those less than 50 Vac/60 Vdc, depending on the standard or term applied and are referred to as Extra-Low Voltage or Safety Extra-Low Voltage (ELV/SELV).

📖 The Low-Voltage Directive applies to all safety aspects of electrical equipment, including protection from mechanical and other hazards.

As a designer you should first give careful consideration to the selection of the components and construction requirements, which minimizes the risk that the product will fail its first test, thus, causing costly product redesign. Don't rely on test results to be the stimulus for the safe design of the product; consider safety before addressing other design elements.

📖 Design and build a product to meet the standards. Enclosures and tests ("box-it and pot-it") are never substitutes for a safe design and should *not* cover for noncompliant components or improper construction.

A product's conformity to the European safety standards relies on the use of proper *component* and *construction* principles. Testing is performed *after* a sound design is in place.

The Hierarchy of Conformity Checks:

1. Check for *components* with European "Approval Marks" as positive evidence of compliance ("CE" *not* considered).
2. Evaluate the products *construction* according to the relevant standards.
3. Perform all relevant *tests*, after components and construction are in order.

📖 In only rare cases should safety testing come first. If testing is performed prior to component and construction assessment and deficiencies are discovered, safety and/or EMC retesting may be necessary.

Principles of Safety

To meet the relevant safety standards and manufacture safe products, it is essential that designers understand the principles of safety. The following is not an alternative to the detailed requirements of the standards, but is intended to provide designers with an appreciation of the principles on which these requirements are based. The requirements of safety standards are, in general, intended to prevent injury or damage caused by the following hazards:

- Electric shock
- Energy
- Fire
- Heat
- Mechanical
- Radiation
- Chemical
- Materials
- Others

Some key terms and definitions are essential in understanding the European safety philosophy.

Electric shock is caused by current passing through the human body. Even currents in the milliamp region can cause a reaction in people, resulting in injury from an involuntary reaction, such as a quick movement that can cause a person to fall or to strike a secondary object. Higher currents may cause more damage. Voltages less than 40 V peak or 60 Vdc are not regarded as dangerous assuming that parts that users can touch are properly grounded or insulated. Electric shock can occur when there is a breakdown of insulation between parts of normally "hazardous voltage" and "SELV" circuits or accessible conductive parts.

SELV, Safety Extra-Low Voltage, is a safe secondary circuit that may be touched by the user (i.e., connector I/Os, wiring for peripherals, some exposed PCB traces). SELV circuits are so designed and protected that under normal and single-fault conditions, their voltage does not exceed a safe value such as 42.4 V peak or 60 Vdc, depending on the standard applied. Hazardous voltages (> 50 Vac/60 Vdc) and SELV circuits (user touchable) must be properly segregated (distance) or separated by earthed screens (metal) or isolated by double/reinforced insulation (two to three layers of insulation). *Reinforced insulation* is required between hazardous voltages and SELV and is comprised of *basic* and *supplementary insulation*. Supplementary is an independent insulation applied in addition to basic (e.g., two levels of protection).

Energy hazards may exist from the outputs of high current power supplies. Injury may result in burns or molten metal by short circuits between adjacent poles such as by metal or ring on finger bridging the supply outputs. The hazard may still exist even in low-voltage circuits where the current is high. Two levels of protection shall be provided such as by insulation, grounding, shielding, or safety interlocks. Operators must be protected from shock and energy hazards exceeding these limits; 42.4 V peak, 8 amps, 240 VA, and 20 joules energy.

Operators are all the persons who use the product, excepting service personnel. Safety requirements assume that operators do not mean to create a hazardous situation and are oblivious to electrical and other hazards. You also must assume that the operator does not normally possess tools reserved for service personnel and maintenance purposes. Equipment must protect janitors and casual visitors as well as operators. An *operator access area* is any area that, under normal operating conditions allows access without the aid of a tool, for example, by a person's hand or fingers alone. Opening a hinged door by hand, without a tool, makes the area behind the door an operator access area, and all hazards shall be adequately guarded or the door interlocked to remove hazards before access. A *tool* is reserved for service personnel and defined as any object that can be used to operate a screw, latch, or similar fixing means.

Service personnel have training and experience and should be aware of potential hazards while performing a task and can take measures to minimize the danger to themselves and others. Service personnel have access to maintenance areas and will be reasonably careful in dealing with the obvious hazards. The product's design, however, should protect service personnel from any possible mishap by using shields for hazardous voltages, segregation of SELV from hazardous voltages, warning labels, and interlocks. A design should protect service personnel from unexpected hazards, such as accidental or inadvertent touching of live electrical components and other hazardous parts during servicing. The *service access area* is an area, other than the operator access area, that service personnel can access even with the equipment power on.

Safety interlocks, common to machinery, provide a means either of preventing operator access to a hazardous area until the hazard is removed or of automatically removing the hazardous condition (i.e., electric shock, moving parts) when access is gained. Safety interlocks have special requirements, such as fail-safe design, positive opening, and nonoverridable type.

Double fixing is a method of ensuring adequate retention of wire terminations and protecting the user should a single fault occur. According to the *single-fault* concept, two levels of protection must be provided for operators to prevent electric shock caused by a fault such as a wire breaking, fastener or connector coming loose, and other malfunctions. Various methods of double fixing should be provided as protective measures, such as liberal use of wire ties, double crimp terminations and sleeving so that a single fault and its resulting faults will not create a hazard.

Other hazards such as fire, mechanical, heat, radiation, and chemical reaction, must also be prevented to protect the operator and service personnel even if the main product standard does not specifically refer to those hazards. Refer to the appropriate standards for requirements on the limiting, guarding, and warnings for these hazards.

Instructions, warnings, and symbols may be required to alert users and service personnel to possible hazards. The type, size, color, and symbols are specified in the standards and sometimes mentioned in a directive (machinery). For electrical products the standards dictate the necessary warnings and where they must be provided (product and/or manual). The standards state that any required safety instructions and markings be in the language of the country where the product is used. And machine manufacturers must supply user manuals in that language. As specified in the standards, suitable warnings and symbols shall be provided for hazards, but warnings shall not take the place of a safe design. Protection of the *operator* from "all possible hazards" must be ensured by two levels of protection; protection of *service personnel* against "unexpected hazards" is achieved with one level of protection.

Design Guidance

I have attempted to simplify the concepts so that technicians, engineers, and other professionals in the industry can understand and apply them. I have numbered the design tips and topics for easy cross-reference to the Safety Checklist at the end of this section. Remember that there is no substitute for experience and proper training and that the design tips presented here do not take the place of the standards. The design tips and subjects discussed are:

1. Component Safety and Liability
2. Component Selection and Conformity Verification
3. Constructional Data Forms (CDF)
4. Equipment Classification
5. Power Consumption
6. Rating Label
7. Markings and Indicators
8. The Single-Fault Concept
9. Input, Wiring, Terminations, and Identification
10. Insulation and Separation of Circuits
11. Grounding
12. Power Disconnect
13. Circuit and Thermal Protection
14. Stability and Mechanical Hazards
15. Warnings, Instructions, and Languages
16. Flammability of Materials
17. Electrical Safety Testing
18. Production Tests
19. Additional Requirements for Machinery:
    A. Protective Measures
    B. Mains Disconnect Switches
    C. Emergency Stop Switches
    D. Fault-Tolerant Components and Safety Circuits
    E. Transformers
    F. Motors
    G. Wiring
    H. Protective Earth
    I. Access Areas
    J. Enclosures
    K. Functional Markings
    L. Item Designations
    M. Warning Symbols
    N. Manuals
    O. Technical Documentation

1. Component Safety and Liability

Any item that is used in the composition of, or intended to be built into, end products or machines, is called a *component*. *Products* are stand-alone equipment that are comprised of components and are ready to use by an operator.

With regard to equipment safety, there are two types of safety sensitive components:

1. *Critical components* are any components that may influence the safety of a product, such as those that operate at mains supply (120/230/400 Vac) or hazardous voltages (> 50 Vac or 60 Vdc). Examples of critical components are inlets, filters, switches, motors, circuit breakers, power supplies, and transformers. Components that may operate at lower voltages (i.e., 12 or 24 V) and may affect safety are also considered critical components and examples include emergency stop switches, door interlocks, relays, secondary fuses, thermal cutouts, fans, and sensors.
2. *Safety components,* as opposed to *critical components,* fulfill a specific safety function when in use and the failure or malfunctioning of the device places exposed persons in imminent danger. Safety components are regulated by the Machinery Directive and examples include light curtains, two-hand controls, and sensor mats.

Under the Low-Voltage Directive (LVD) the CE marking is affixed primarily to finished products that (1) are ready to use and (2) can operate. Examples include electrical equipment, appliances, apparatus, and systems that are intended for the final user and placed on the market as a single commercial unit. Components designed for use within (built-in) products or machines must comply with component safety standards to satisfy the General Product Safety and Product Liability Directives, but do not require CE marking because they have no autonomous use (e.g., they do not "operate" within the specified voltage range and are not "ready-to-use").

✓ *Notice.* Components to be incorporated into equipment, where their safety depends to a very large extent on how they are integrated into the final product or machine, *cannot* be CE marked!

Examples of components that do *not* require CE marking are electromechanical components (relays, microswitches, connectors), active components (transistors, diodes, opto's, ICs), and passive components (capacitors, coils, resistors).

The European Commission's *Guidelines on the Application for the Low-Voltage Directive* clarifies the issue of CE marking for components, stating;

> In general, the scope of the Directive includes both electrical equipment intended for incorporation into other equipment and equipment intended to be used directly without being incorporated. However, some types of

electrical devices, designed and manufactured for being used as basic components to be incorporated into other electrical equipment, are such that their safety to a very large extent depends on how they are integrated into the final product and the overall characteristics of the final product. These basic components include electronic and other components. Taking into account the objectives of the "Low Voltage" Directive, such as basic components, the safety of which can only, to a very large extent, be assessed taking into account *how* they are incorporated, are *not* covered as such by the Directive. In particular, they must *not* be CE marked.

The Commission explains that only products and components that are "ready to use" and actually (de facto) "operate" may bear CE marking under the LVD. There are a few cases, however, where CE marking is appropriate for components, such as for those intended for 'building installations' which include; household switches, lamps, starters, wires, plugs, fuses, and some types of motors and transformers.

The EMC directive allows CE marking of components that (1) have an "intrinsic function" and (2) are "sold to the end user," but not when sold to manufacturer's or assemblers. Some components may bear CE marking under the EMC directive. Examples of components include personal computer circuit cards, computer disk drives, PLCs, stand-alone power supplies, electric motors (except induction), and electronic temperature controls.

Because of the competitive nature of the marketplace, component suppliers often place CE on components for sales and marketing purposes, even when a directive does not specifically allow it. There are a few provisions for CE marking of components under the Low-Voltage and EMC directives that some manufacturers and suppliers take advantage of to satisfy customers' demands (albeit oftentimes misinformed) by affixing an *unqualified* CE marking. Manufacturers' CE markings and self-declarations for components, without EU type-approval marks, may make the claim of conformity difficult to prove. Most components do not need CE marking, and where it exists "CE" is *not* evidence of compliance from a European recognized third party.

⬤※ *Warning!* The Commission has disallowed CE marking of components when "their safety to a very large extent depends on how they are integrated into the final product." Some suppliers still persist in marking components with an *unqualified* CE.

A major problem with products and machines submitted to EU-testing bodies is that they fail assessment because the components do not conform to the relevant standards. Failure is most likely when nonapproved components are used. Since the 1970s (Old Approach) a self-declaration according to the LVD was possible, but most components carrying only a manufacturer's declaration failed when tested by a European body. Under the New Approach nothing has changed. Safety and critical components still must comply with the standards, but those that have only the manufacturer's self-declaration and CE marking and lack type-approval markings usually fail testing by a EU body.

💣 Because of misinterpretation or misapplication of the standards, nonconformity of component safety is commonplace, even when tested by the component manufacturer according to an EN or IEC standard. *Most* nonapproved components *fail* safety tests by European testing and certification bodies.

The equipment manufacturer takes complete responsibility for the end product or machine and the components it specifies. If no positive evidence exists (CE *not* considered), it is always prudent to perform all relevant component tests or select a different component so no questions arise (see Figure 6-1). Safety testing and certification of components should rest with the component manufacturer and not burden the company that purchases them, instead leaving it up to the product designer to only verify through "positive evidence" that the component complies with the relevant standard (see Figure 6-2).

❗ It is the end product or machine manufacturer who takes conformity responsibility for the equipment's design *and* the "components" they specify.

Utilizing nonapproved components in the equipment places the end-product manufacturer in a precarious position especially with regard to testing, documentation, and ongoing conformity of the components. The CE marking is only a self-declaration, and if the component has no European type-approval mark, the end-product manufacturer may have to do additional testing:

1. Testing nonapproved components adds cost, time, and risk since *the component often fails the assessment,* even when a CE marking or other non-European mark (UL/CSA) are present. Remember, manufacturer's self-declarations for products and components have been allowed under the LVD since 1973 and not generally recognized as evidence of compliance by European testing bodies or enforcement authorities.
2. If passing results are obtained in the end equipment for a nonapproved component, it may then be used in that model only, restricting the component as *application sensitive*. Retesting may be required for the same component in other models.
3. Using nonapproved components makes the *product manufacturer responsible*. This includes responsibility for the component's design, construction, documentation, and ongoing compliance, areas where the end-product or machine manufacturer have no real control.

📖 Specifying European type-approved and marked components proves to be the path of least resistance and the lowest-cost alternative.

To ensure safety compliance and limit testing, end-product manufacturers often demand *prequalified* safety-sensitive components with EU type-approval marks (VDE/TUV). These component marks help to ensure compliance with standards and reduce testing. Components with VDE/TUV marks need only be checked for proper application and use in the product. European-approved components may cost a little more, but they are usually more reliable, and in the long run will save the product designer much aggravation and time. Figure 6-1 shows some of the standards that require components safety compliance.

EN 60950 (IEC 950)—Safety of Information Technology Equipment

- Where safety is involved, components shall comply either with the requirements of this standard or with the relevant IEC component standards. (1.5.1)
- Evaluation and testing of components shall be carried out as follows:
  —a component certified by a recognized testing authority for compliance with a standard harmonized with the relevant IEC component standard shall be checked for correct application and use in accordance with its rating. It shall be checked to the applicable tests of this standard as part of the equipment with the exception of those tests which are part of the relevant IEC component standard;
  — a component which is not certified for compliance with the relevant standard as above shall be checked for correct application and use in accordance with its specified rating. It shall be subjected to the applicable tests of this standard, as part of the equipment, and to the applicable tests of the component standard, under the conditions occurring in the equipment.
  *Note.*The applicable test for compliance with a component standard is, in general, carried out separately. The number of test samples is, in general, the same as that required in the component standard. (1.5.2)

EN 61010-1 (IEC 1010-1)—Safety of Test and Measurement Devices in Lab Use

- Where safety is involved, components shall comply with the applicable safety requirements specified in relevant IEC standards
  If components are marked with their operating characteristics the conditions under which they are used in the equipment shall be in accordance with these markings, unless a specific exception is made. (14.1)

EN 60335 (IEC 335)—Safety of Electronic Products for Household and Similar Use

- Components shall comply with safety requirements specified in the relevant IEC standards as far as they reasonably apply. (24.1)

EN 60204-1 (IEC 204-1)—Safety of Machinery, Electrical Equipment of Machines

- Electrical components and devices shall be suitable for their intended use, e.g., industrial (heavy, light), commercial, leisure, domestic, and shall comply with the relevant European Standards [EN] where such exist. In absence of European Standards, compliance shall be to available International Standards [IEC]. (4.2)

FIGURE 6-1   Product Standards Require Component Safety Compliance

2. Component Selection and Conformity Verification

If a critical or safety component does not have the proper approval mark, the test and certification body may fail the product or require testing of the component. Manufacturers doing a self-assessment of their equipment should use the same criteria as the EU bodies and select only components that they know have been tested, certified, and bear an EU approval mark. Anything less may require additional testing and/or review by a European component expert. See Figure 6-2 for component acceptance criteria.

Because of the almost universal acceptance of the European standards (EN and IEC) around the world many suppliers now offer components that are EU type-approved, in addition to the U.S. recognition. The U.S. is also moving toward acceptance of the European product standards. Unfortunately, at this time U.S. and EU product and component standards are different. The United States focuses on fire hazards and materials, whereas Europe stresses shock hazards and construction. Because of these and other substantial safety differences, component suppliers realize the need for dual approvals to satisfy both the U.S. and International requirements. As a minimum, one mark for North America (UL) and one for Europe (VDE) should be considered. Many component suppliers now offer components with this dual certification. A "certificate and test report" from the testing and certification body supports the type approval mark. The mark is affixed to the component and is visually recognized by interested parties as positive evidence of compliance.

✓   Learn to recognize EU approval marks. Look at a component and identify European approval marks, such as:

Know and recognize the many European approval marks on components. These symbols and logos from EU certification bodies give the consumer the added assurance they demand. Also be able to differentiate "approval marks" from "self-declarations." As mentioned, there is no such thing as an "IEC" or "CE Approval." Approvals and certifications are granted only by European accredited testing and certification bodies. The alpha characters IEC and VDE, sometimes found on components and in marketing literature, are a reference to the standard that the component supplier claims to meet and are not independent EU third-party approvals. Do no confuse the self-declaration markings (alpha characters, not symbols) of "IEC" or "VDE" or "CE" on the component or in a sales brochure for a "type-approval mark." If in doubt, contact a European body for verification. The following are standard numbers or prefixes and are only manufacturer's "self-declarations."

Rating	Evidence of Conformity	Conformity Status (1) and Actions Required (2)		Verdict
A	"EU Type-Approval Mark"	1	Component was tested by European accredited testing and certification body and bears the "Type-Approval Mark" as positive evidence on conformity. Test report is on file with EU body and available on request.	Pass
		2	End-product manufacturer confirms approval by observing "Type-Approval Mark" on component and obtains copy of the approval certificate, if necessary, for their technical file.	
B	"Accredited Lab" tested	1	Occasionally components are tested by a EU "Accredited Lab" (DAR for Germany) and do not bear a type-approval mark. The "lab accreditation certificate, test report, and test verification" must be readily available.	Conditional acceptance (case by case)
		2	The lab "accreditation certificate" and "test verification" is obtained by end-product manufacturer and placed into the technical file. It is also advisable to obtain a copy of the "test report" for review.	
C	"Manufacturer Self -Test"* ("CE" marking)	1	Component supplier claims to meet EU safety standards (i.e., EN, IEC, VDE) and offers "CE" marking in lieu of type-approval mark.	Questionable (reject or test)
		2	Since self-declarations and "CE" are *not* positive evidence, the end-product manufacturer should, in order of preference: (a) reject component and select alternative, or (b) test component, if pass document,* or (c) if EU "type-approval" exists on similar components within same series, expert reviews test report and sample component for acceptance/rejection.*	
D	"Evidence not available"*	1	No positive evidence available (e.g., no EU approval mark or third party report).	Fail
		2	Product manufacturer rejects component, or performs complete testing, according to the relevant standard(s). Testing by EU third-party expert preferred. *	

* Note: The component manufacturers declaration or CE marking, if present, should not be used to verify conformity. Lower ratings indicate a increased risk of non-conformity or potential failure when tested. If the component in question falls within ratings C or D the component and test report should be assessed by a European safety expert prior to its acceptance. In the case of testing non-approved components in the end-product, and if successful test results are achieved, the end product manufacturer takes on the conformity responsibility of the component and guarantees its on-going compliance.

© Lohbeck

FIGURE 6-2 Component Acceptance Criteria—"Evidence of Compliance"

! These are *not* approvals from independent third parties, even when marked on the component: CE, IEC, VDE, IEC 947, VDE 0660, to name just a few!

For components and products that successfully pass testing, a certificate is issued by the EU body giving the manufacturer the right to affix the approval mark. In addition to the mark on the component a copy of the *certificate* may be needed to verify the component's ratings and part number and to bring to light any "restrictions for use." Some designers and buyers are satisfied if they see the approval mark on the component, but it's a good idea to obtain the approval certificate for review and to place in the technical file. Always request the approval certificate for complex components (such as monitors and power supplies) since restrictions may be written on the certificate or in the installation instructions, which must be checked for suitability in the end application. Some examples of restrictions that may be found on a power supply certificate are:

- For use with external fuse; rating of . . .
- For use with forced air cooling of . . .
- For vertical mounting only . . .
- Isolation system . . . "hazardous energy" outputs.

It is important to request the approval certificate to answer all your questions. Once you have the certificate you should also verify that:

- The components part number matches the component number on the certificate;
- The voltage, current, and other ratings are acceptable for the intended use;
- The appropriate standards have been applied; and
- Any restrictions for use.

3. Constructional Data Forms (CDF)

The critical component parts list, or *Constructional Data Form (CDF),* is an important tool to aid in component selection, their application criteria, and verification of conformity marks. Figure 6-4 shows the CDF. It is also extremely useful during the design, purchasing, and manufacturing control of components used in the equipment. The CDF is a listing of all safety-sensitive components (see Figure 6-3 for component types) with the relevant technical data such as (1) component description, (2) manufacturer's name and part number, (3) rating information, and (4) EU type-approval marks found on the component. From the CDF anyone can see at a glance which are the critical components for safety and most important it identifies the European type-approval mark(s) found on the components (right column of CDF).

Once the end product is in conformity with the relevant standards, the CDF can be used by engineering, purchasing, and manufacturing to control the parts and ensure that the company uses only approved replacements or alternates. Alternate or additional components can be easily added during design on the initial CDF or at a

Product and Machine Components:	Machinery Specific:
• Inlets and outlets • Plugs and connectors • Power cords • Strain reliefs • RFI filters, chokes, caps • Circuit breakers • Fuses (user accessible) • Fuseholders • Mains disconnect switch • Power supplies • External supplies • Fans (AC and DC) • Terminal blocks • Relays • Disk drives (all types) • Thermal cutouts • Current-limiting devices • Voltage select switch • CRTs, monitors, etc. • Transformers** • Motors** • Printed circuit boards*** • Plastics and enclosures*** • Conductive coatings*** • Wire insulation*** • Air filters*** • Batteries*** • Others components >50 Vac/60 Vdc • Components in safety circuit (regardless of voltage, i.e., 12/24 Vdc operation)	• Enclosures (for elec. controls) • E-stop switch • Service disconnect device • Interlock switches • Position sensors • Contactors • Relays (ind. types) • Circuit breakers (ind. types) • Fuses (internal and external) • Function controls/switches • Push-buttons and switches • Indicator lights and towers • Heaters and elements • Solenoid valves • Alarms (audio and visual) • Robots and controls • PLCs, drives, and controllers • Flat panel displays • Products; PCs, VDUs, UPS, air conditioners, lighting, etc. • Safety components **** • Others

* "CE" markings and declarations are not considered as evidence of conformity. Components bearing the appropriate European "type-approval mark" (VDE/TUV) need only to be checked for proper application in the end product. Components without an approval should be assessed and tested according to the relevant EN/IEC component standard and for application in the end product.

** Type-approved transformers or motors are in some cases difficult to source, especially the larger types, but in any case must comply to the relevant EN/IEC component standards.

*** UL/CSA flammability classification may be acceptable. Refer to the standards or consult a EU body for the requirements regarding flammability ratings (i. e., 94V-X), battery types, protection circuits, and testing.

**** Safety components such as light curtains, sensor mats, and two-hand controls are regulated by Machinery Directive 89/392/EEC which requires a manufacturer's declaration of conformity but no CE marking. EU "type-approval marks" are also highly recommended for safety components.

© Lohbeck

FIGURE 6-3 Components Requiring Evidence of Conformity: "Type-Approval Marks"*

TUV Rheinland of North America, Inc.	Certificate No:	File No:	Attachment No:	△ TÜV Rheinland
	S98XXXXX	141-DL/tl/E97XXXXX.01	1	
		(to be filled in by TUV Rheinland of North America, Inc.)		

Constructional Data Form				page 1 of X
Applicant/Manufacturer:	XYX Corp.		Rated Voltage:	AC 115/230 V, 60/50Hz
Type of Equipment:	Personal Computer		Rated Input:	4/2 A
Type or Model No. :	RMS-XY (X=0-9, Y=A-Z*)		Protection Class:	☒ Class I, ☐ Class II, ☐ Class III

Built in critical components: (switches, inlets, capacitors, filters, heaters, motors, transformers, power supplies, protection devices, etc.)
Complete table below, with EU approval mark(s) in the right column (CE and IEC not valid). UL/CSA considered for plastics flammability only:

kind of component	manufacturer and part number	information about type, current, power, etc.	approval mark

(signature and date of TUV engineer) **TUV Rheinland of North America, Inc.**	(city, state) (date)
	(signature of applicant/manufacturer)

NOTE:. TUV-CDF modified for clarity. CDF formats may vary.

FIGURE 6-4   Constructional Data Form

later date by simply adding additional pages to the CDF. This product change is known as an "alternate construction." If the alternate or replacement part is simple and has similar specifications and an EU type-approval mark, the change may involve only paperwork, depending on the component and its application. Substitution of more complex components such as transformers or power supplies may require additional testing in your intended application, again assuming that the component bears an EU approval mark as a minimum. The CDF is especially helpful for factory checks at incoming inspection to verify the components approval marking before accepting them into stock. The CDF can also be referred to during production to verify that manufacturing has assembled the proper components into the product and allows inspectors an easy means of checking for the proper component, number, or approval mark. I recommend listing the components in order starting from the input (inlet or cord) and working your way through the product's circuit diagram. Components at line voltage (120/230/400 Vac) and at hazardous voltage or energy levels (>50 Vac/60 Vdc) must be listed and controlled. Components below these limits are not usually listed unless they have a safety function, such as a door interlock or safety switch operating at any voltage. Components below the safe voltage limits with moving parts or made of hazardous materials, such as DC fans, disk drives, and batteries, should, however, be listed. Do not list any components that are not safety relevant such as I/Cs and most components in SELV circuits.

Plastics used in the construction of a product must meet the relevant flammability ratings, such as V-2 for PCBs and V-1 for enclosures. In this case, a plastic that has been tested and rated by a North American agency (UL/CSA) according to the relevant flammability requirements of the standards may be acceptable. Consult an EU testing body for their acceptance and documentation criteria (see Figure 6-4 for CDF and Figure 6-3 for critical components examples).

4. Equipment Classification

The *electric shock classification* of the equipment will determine the extent of insulation needed to protect users and service personnel.

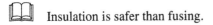 Insulation is safer than fusing.

Insulation is achieved by separating circuits and is required, for example, between user-accessible parts and live parts via insulation layers, thickness, and/or distance (creepage and clearance). There are three classes of equipment, Class I, II and III, with Class I and II products generally covered by the Low-Voltage Directive:

- *Class I* equipment utilizes earth ground and are typically mains operated (i.e., 115/230/400 Vac, 3–5 wires) electrical products such as desktop computers, test and measurement devices, machines, or stationary appliances.
- *Class II* products are ungrounded and rely on protective earthing (2 wires). Examples include VCRs, TVs, power tools, portable radios, and other handheld or portable appliances.
- *Class III* products are generally considered electrically safe since they operate outside the specified voltage range of the LVD (< 50 Vac/75 Vdc). Battery-operated products operating outside the specified range are not covered by the LVD. Products which generate internal high voltages are also excluded from the LVD's scope provided the high voltages are not accessible via sockets or other accessible parts. The LVD is applicable, however, to battery-operated Class III devices, if they can be operated with a mains-connected power supply or charger (e.g., a laptop computer).

The *mobility classification* of the equipment is another important factor affecting the products testing and construction requirements, such as impact and drop tests, leakage limits, enclosure strength, and labeling. The equipment categories are:

1. *Handheld equipment.* Products intended to be held in the user's hand during normal use.
2. *Movable equipment.* Equipment that is either 18 kg (39.5 lb) or less in mass and not fixed or that has wheels, castors, or other means to allow movement by the operator as required to perform its intended use.
3. *Stationary equipment.* Equipment that is *not* movable equipment.
4. *Fixed equipment.* Stationary equipment fastened or secured at a specific location.
5. *Built-in equipment.* Equipment that is intended to be installed in a prepared recess, such as a wall. *Note:* Built-in equipment may not have an enclosure on all sides, as some sides may be protected after installation.
6. *Direct plug-in equipment.* Products intended to be used without a power supply cord; the mains plug is an integral part of the product's enclosure, and the weight of the product is supported by a socket-outlet.

5. Power Consumption

It is important to know the equipment's actual power requirements to design the markings and rating labels. Calculating the equipment's input is often unreliable. Through measurements, establish the equipment's power input under normal operating conditions and with all possible loads applied. The input is verified by measuring the input current (amperage) to the product. Measure and record the product's input power at the desired operating voltage, which is typically 230 Vac (or 400 Vac for three phase). The old voltages of 220 to 240 Vac (or 380/415 Vac) may still apply in some countries. Please note that 208 Vac, common in the United States, is not generally available in Europe. For ITE, take the measurements at the desired frequency (50 Hz) and at ±10% for a singular voltage rating (230 Vac) or +6 to −10% for a voltage range (220 to 240 Vac). Because input current varies at the extremes of the test range, it may reduce testing if the rated voltage (shown on nameplate) is limited since all testing will be performed at ±10% or +6 to −10% of the rating value (see product standard for test ranges). For example, if a computer is rated at 200 to 240 Vac, the safety tests would be performed at +6 to −10% of the rating shown on the label; testing range of 180 to 254 Vac. A printer with a single rating of 230 Vac would require ±10% of 230 V for a test range of 207 to 254 Vac.

After taking the input measurements, select an input rating for the label slightly higher than the measured value such as measured = 2.75 A, label marking = 3.0 A. Do not select an input for the label that is well above the measured value because all product testing would then be performed at that value by adding loads, and so on. The products operating input power may not exceed 110% of the rating shown on the label.

6. Rating Label

Products shall be provided with a input rating marking (label or plate) to specify the input voltage, current, and frequency. The label must also indicate the manufacturer's name and type number (model number) and should be located adjacent to the inlet or power entry. Additional information, such as serial numbers, date codes, and approval marks may be on the same label. The label should be on the exterior of the product and easily recognizable by the user for portable equipment or by the installer for larger equipment. If the rating label is located behind an operator-accessible door (not recommended), for example, within ITE, a visible temporary marking should also be used. Do not use the words *voltage, amps, hertz* or others when IEC symbols exist. Use proper European symbols whenever possible such as, V for voltage, A for current, and Hz for hertz. A dash (–) is used to indicate a range (220–240 Vac) and a slash (/) for either or (120/230 Vac).

If the equipment utilizes only two wires for input (line and neutral), as a double insulated product, and there is no ground connection, it is then necessary to mark the product with the Class II symbol. Examples of rating labels for single phase products are:

XYZ Corp.	XYZ Corp.	XYZ Corp.
Model No. 1234	Model No. 1235	Model No. 1236
115/230 Vac	230V, 5A, 50/60 Hz	220–240 Vac
3 A, 50/60 Hz		2.0 A, 50 Hz

Using AC is optional for ITE. The symbol " $\sim$ " may be used instead of AC.

Examples of rating labels for three-phase equipment are:

ABC Machine Inc.	ABC Machine Inc.	ABC Machine Inc.
Model XXXX	Model XXXX	Model XXXX
3/N AC 400 V	3/PEN AC 400 V	3/N/PE AC 400 V
50 A, 50 Hz	32 A, 50 Hz	125 A, 50 Hz
(Four-wire, with neutral)	(Four-wire, with neutral and protective function)	(Five-wire, with neutral and protective earth)

7. Markings and Indicators

Manual control devices shall be clearly and permanently designated with regard to their functions on or adjacent to the actuator, such as the OFF symbol O and

ON symbol I. If only a part of the product is switched off, the standby symbol ⊕ may be used. Numerous other symbols exist for switch and control indications, such as for push-push or hold-to-run. Figure 6-5 shows some examples of symbols.

Fuses, whether internal or external (accessible) shall be marked with fuse type and rating adjacent to the fuse. The fuse symbol is also recommended but not mandatory. Use only IEC symbols for the fuse ratings and type (F = fast acting, T = time delay, M = medium time delay). A table listing the fuse number and rating may be used where several fuses are found in one location. Examples of fuse markings are:

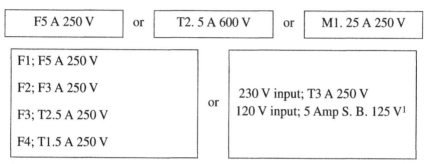

| F5 A 250 V | or | T2. 5 A 600 V | or | M1. 25 A 250 V |

F1; F5 A 250 V		
F2; F3 A 250 V	or	230 V input; T3 A 250 V
F3; T2.5 A 250 V		120 V input; 5 Amp S. B. 125 V[1]
F4; T1.5 A 250 V		

[1] Fuse rating for U.S. products only.

Outlets, (internal or external) also need the voltage and maximum power markings next to the outlet, such as:

2 A 250 V

Maximum

Use "functional markings" such as ON/OFF, START/STOP symbols for all controls and indicators whenever possible, unless their function is obvious. Proper markings and colors are important especially if the control or indicator in question is related to safety. Green and red are controlled indicator colors, with a *red* light used to indicate an *Emergency* condition that instructs the user to disconnect the power to the equipment. A *green* light shows a normal power, run, or ON condition. *Yellow* is a warning color used to indicate an impending hazardous condition (see EN 60073). Some standards may permit alternate colors, such as in coffee makers or film processing products. Most safety standards cover the specific information required for safety functions and markings.

8. The Single-Fault Concept

The European safety standards stress an important principle called the *single-fault concept* that is at the core of safety philosophy. This principle states that even under a single fault (a wire coming loose, component failure, short/open circuit, insulation failure, etc.) at least one level of protection (basic insulation) must be

Symbol	Publication	Description
⎓	IEC 417-5031	Direct Current
∿	IEC 417-5032	Alternating Current
⏛	IEC 417-5016	Fuse
⏚	IEC 417-5017	Earth Terminal
⏣	IEC 417-5019	Protective Earth Terminal (PE)
⏛	IEC 417-5020	Frame or Chassis Terminal
▽	IEC 417-5021	Equipotentiality
I	IEC 417-5007	On (supply)
O	IEC 417-5008	Off (supply)
⏻	IEC 417-5009	Stand-by
⏼	IEC 417-5010	Push on - Push off, alternatively
⏽	IEC 417-5011	Push button, hold to run
◁	IEC 417-5104	Start (of action or operation)
▽	IEC 417-5110	Stop (of action or operation)
⊙	IEC 417-5264	On (only for part of equipment)
⊙	IEC 417-5265	Off (only for part of equipment)
▣	IEC 417-5172	Equipment protected by Double or Reinforced Insulation
//	ISO 7000-1027	Reset

Note: For additional symbols and more information refer to EN 60417 (IEC 417) and EN 50099. *Symbols courtesy of Hazard Communications Systems, Inc (HCS), Milford, PA.*

FIGURE 6-5 Common Symbols

maintained after a single fault to ensure adequate protection of the user. This concept is sometimes referred to as *double improbability* which means that there is always a second means of protection or insulation should the first one fail. Double or reinforced insulation provides the required protection (i.e., two levels before fault) between live or hazardous parts and the user, therefore, even after a single fault has occurred the operator is still safe (basic; one level after fault).

With the single-fault concept we can determine the protection needed to satisfy the electrical safety standards, such as basic, supplementary, or reinforced insulation. In addition, it is through single-fault analysis that the number and type of abnormal tests are determined. Perform a detailed safety fault analysis to determine which faults can occur to establish the proper insulation type(s) and identify the abnormal testing required. Using European type-approved components and construction techniques (i.e., proper thickness/sleeving and double fixed terminations for internal wires) greatly reduces the fault testing involved. Fault testing of nonapproved components such as power supplies can often cost more than testing the product or machine itself and the component oftentimes fails anyway. The scope of this section is purposely limited and assumes that all critical components bear European-type approval marks. There are other books and standards that the manufacturer can refer to for component assessment and testing (see previous component sections 1 to 3 to limit risks).

Machines safety circuits sometimes require special components such as relays, contactors, interlocks, and E-stops. Common terms associated with these machine components are *control reliable, fault tolerant,* and *fail-safe,* which means that they fail to a safe condition after a single fault (not multiple faults).

9. Input, Wiring, Terminations, and Identification

All wire sizes and insulation must be suitable for their intended use and rating. Since wiring is oftentimes critical to safety, it should have the proper classification, such as a V-2 flame rating for insulation on internal wiring or <HAR> marked for external fixed power cords. English or metric sized wires are acceptable.

Internal wires are held in place independently of the connection, to meet the single-fault requirement, by wire ties or similar methods. Make the second fixing as close to the initial connection point as possible so that if the wire breaks or comes loose, it will not make contact with metal or live parts. It is assumed that two independent fixings will not come loose.

Using the proper IEC type terminals as instructed by the terminal manufacturer will reduce any termination problems. Wires are not considered reliably secured to terminals unless there is either an additional fixing provided near the terminal or the terminal has terminators (i.e., ring lugs). For press-on or similar terminators, a double crimp is preferred (e.g., one crimp on the wire and one on the insulation).

Because of the many plug and socket types in Europe, designers often prefer inlets for use with detachable cord-sets instead of fixed cords and plugs. The cord-set does not need to be listed on the CDF. EN 60320 type inlets are available for up to 15 A input, and EN 60309 pin and sleeve types are available through 150 A or higher. The U.S. NEMA twist-lock style inlet does not meet the IEC standards and, therefore, is not allowed for use in operator-accessible areas. Inlets or inlet-filter modules are generally easier to deal with for the equipment manufacturer and the installer than fixed cords. Tabs are typically provided on the internal side of the inlet for solder or press-on connectors. Loop the wire through the hole of the tab, prior to soldering, and tie or sleeve for the second fixing.

Fixed power cords are common to machinery and household appliances, but care must be taken to ensure that the cord, strain relief, plug, and terminal block are type-approved and that the proper termination methods and makings are utilized. IEC-type terminal blocks (touch safe) are the preferred method for termination of the incoming power wiring. Other methods of termination are possible, such as for connection directly to circuit breakers, mains disconnect switch, PCBs and tabs/screws on filters. Refer to the product standard for the alternative methods of input terminations. Solder connections shall not be the only method of fixing and in some cases are not allowed for high current connections.

Terminal blocks and stranded wire insertion pose special problems because standards require a loose strand test to determine if a single strand can touch any conductive part upon insertion into a terminal block. Using the proper terminal block and/or insulation barrier solves this problem. Because of solder cold flow problems tinning should not be used to consolidate the wire ends for terminal block insertion.

Fixed power cords may require replacement several times over the life of the equipment. For cord replacement, the strain relief must allow for a wide range of cord sizes and stay with the equipment during the replacement. U.S.-type strain reliefs typically do not meet IEC requirements. IEC-style strain reliefs are typically plastic (not metal) with ferules/blocks for cord compression.

Input wiring terminals must be properly marked, such as for protective earth (PE), neutral (N), and line(s) (L), depending on the power distribution system and standard applied:

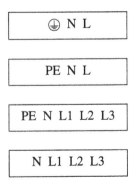

The proper wire colors and approval markings must be used for fixed power cords. The fixed power cord should have a <HAR> mark to ensure it meets the IEC standards for wire size, insulation, and colors (e.g., brown for line, blue for neutral, and green-yellow for protective earth [PE]). Do not use green-yellow for purposes other than safety ground. Do not use for SELV, ELV, low-voltage returns, or EMC/RFI grounds. Also, light blue signifies the neutral conductor. In general, for some products covered by the Low-Voltage Directive wires can be any color with green/yellow reserved for safety grounds. You should, however, follow the accepted color conventions throughout. When dressing the fixed input cord wires, make sure to provide an adequate service loop for the PE wire so it would be the last to break if the strain relief fails to hold when the cord is pulled.

10. Insulation and Separation of Circuits

Insulation separates live parts of the product from the user and prevents live parts from coming into contact with each other. For protection against direct and indirect contact with live parts insulation is safer than grounding or fusing. The protective measures are shown in the relative product standards and apply to all products and machines. The types of insulation applied between circuits of different potentials and metal parts depend on the equipment class, possible faults, and the parts in question. The voltage and part or circuit accessibility are also important for proper selection of the insulation type, such as "double or reinforced" insulation between live parts and user-accessible parts. Basic insulation is required between live parts and grounded metal parts.

There are five types of insulation. Basic, supplementary, and reinforced insulation are most often applied for safety purposes.

- *Operational insulation.* Insulation needed for the correct operation of the equipment, such as lacquers or circuit board coatings. Operational insulation does not protect against electrical shock.
- *Basic insulation.* Insulation to provide basic protection (one level) against electrical shock.
- *Supplementary insulation.* An independent insulation applied in addition to basic to ensure protection against electric shock in the event of a failure of the basic insulation (equals two levels).
- *Double insulation.* Insulation comprised of both basic and supplementary insulation (provides two levels).
- *Reinforced insulation.* A single insulation system that provides a degree of protection against electrical shock equivalent to double insulation. It may comprise several layers to meet thickness and/or test requirements.

Separation of the internal wiring, such as between primary and secondary wires or adequate thickness or layers of insulation, is required even within the equipment. Whenever possible it is best to physically separate all primary and secondary wires so that there is no possibility of their coming into contact with each other. When this is not possible, as when primary and secondary wires touch, care must be taken to ensure that the proper insulation level (reinforced) is maintained. As a general rule, when primary and secondary wiring touch, only two layers of insulation is required (one on each wire) if one layer has a minimum thickness of 0.4 mm and the other wires' insulation passes a hi-pot test (1,500 Vac). To obtain reinforced with only two layers of insulation (one on each wire) it is best to use 600 V–rated insulation (≈ 0.7 mm thickness) for the primary wires since 300 V wire insulation has typically less than the required 0.4 mm thickness. Whenever the minimum 0.4 mm thickness requirement is not met, reinforced insulation may be accomplished with the use of three layers of insulation, passing the appropriate hipot test (1,500 Vac for two layers). Either the primary or the secondary wiring may be sleeved with an additional layer, thereby achieving the total of three layers. The primary wires are usually sleeved with solid tubing, such as heat shrink or clear PVC. Porous insulation and spiral wrap do not count as insulation since they typically fail the hi-pot tests. Adequate separation must also be maintained between wiring and exposed circuits such as exposed terminals on components or PCB traces. Secondary wires contacting primary circuits on power supplies or primary wires contacting exposed secondaries are areas that must also meet the reinforced insulation requirements.

Where sleeving is used as supplementary insulation on internal wiring, it must be positively retained in position and is considered as such if it can be removed only by breaking or cutting or if it is clamped at both ends.

11. Grounding

Reliability and proper identification are key to the protective earth (PE) in equipment. Adequate sizing and the proper insulation colors of green with a yellow stripe shall be used for the PE, safety, and chassis ground wires. Clear insulation on braids may also be used. Green/yellow is strictly reserved for the aforementioned grounds and may not be used for low voltage (< 50 V) or EMC/RFI grounds. All accessible metal parts, such as panels and doors, providing guarding from hazardous voltages, must be "adequately earthed" between these parts and the protective earth terminal. Perform a ground continuity test on all user-touchable metal parts at 25 A (12 V source) or 1.5 times the limited input current (1.5 × fuse/breaker rating) depending on the standard. The measured resistance must be less than 0.1 ohm for the ground continuity test.

Masking of paint and ground straps between metal doors and the chassis helps to ensure an adequate safety ground. Grounding through hinges of metal doors is not considered adequate. Ground straps or wires are necessary for all hinged metal doors, that cover hazardous voltages to ensure an adequate safety ground.

One should not be able to disconnect the PE conductor during the servicing operation to replace any parts. This means that the grounding stud or screw should not be used to hold any other replaceable part in place. An isolated ground point is preferred and often required in some standards, along with the protective earth symbol (⏚) for connection of the PE wire only (green-yellow insulation). Only one PE symbol is allowed in the product, unless there is more than one power cord. The chassis ground symbol ⏚ may be used for other safety or chassis grounds. In addition, all ground terminations, such as ring lugs, must be reliably fixed and not allowed to turn. Use a lock washer and nut for each ring lug even when these are stacked on the same ground stud (PE on its own terminal preferred).

12. Power Disconnect

A hand-operated "power disconnect" device shall disconnect the whole equipment from main input supply power. There are several possibilities. A user-accessible "disconnect switch" located on the front of the product is preferred but not the only option. Some of the power disconnect methods are:

1. Disconnect switch (ON/OFF)
2. Circuit breaker
3. Plug-on power supply cord (portable or small equipment < 16 A; user warning required)
4. Service disconnect switch (lockable for machines)
5. Building disconnect device (for permanently connected equipment; warning required)

Mains disconnect devices must meet these requirements: (1) 3 mm minimum contact separation, (2) connected close to incoming supply, and (3) clearly marked with the symbols I for ON and O for OFF.

A double-pole disconnect device (switch, breaker, other) should be used and should disconnect both poles simultaneously. In some products a single-pole device may be allowed to disconnect only the phase conductor when the neutral can be verified, but this is often not possible since plugs are reversible in many EU countries, such as in Germany. Parts that remain live after the disconnect device is opened (OFF) shall be guarded against accidental contact by service personnel and a warning label is necessary adjacent to the hazard areas.

Emergency stop (E-stop) switches, common to machinery and some products, use red mushroom buttons on a yellow background, which is universally recognized by operators and service personnel. E-stop devices have special requirements such as, positive opening, manual reset, fail-safe, and should be type-approved as E-stops (see Emergency Stop Switches section).

13. Circuit and Thermal Protection

Circuit protection devices, such as fuses, breakers, or fault-interrupters, may be required in case of excessive current draws as a result of a short circuit, overcurrent, or earth fault. Several options exist with circuit breakers preferred over fuses. In some of the newer devices several functions are combined into one, thereby reducing the total number of components. An example of this is a combined power-switch/circuit-breaker. The switch/breaker is used as a switch and breaker and senses each line and opens all lines, except the grounded line, simultaneously when a fault current is present. Also, the switch/breaker looks and functions as a standard power ON/OFF rocker switch. In some standards simultaneous interruption of the phase and neutral is required, thereby, precluding the use of fuses.

Fuses must be properly rated for performance in respect to voltage, current characteristics, and breaking capacity, and they must comply with the European standards. Using fuses and fuseholders that are European approved ensures conformity with the standards. A time-delay fuse, type T, may be used to avoid nuisance tripping, but the delay may not result in a fire or overheat condition during a fault test. Select the lowest value possible according to the actual current (*not calculated current*).

The rated current of fuses and other protection devices shall be selected "as low as possible" based on the actual measured input.

In general, for information technology equipment (ITE), user-accessible fuses shall be of a European type (IEC; 5 × 20 mm), whereas, internal fuses may be U.S. sized (3 AG; 1/4 × 1-1/4 in) or an IEC type. Other standards, such as for machinery, require that all fuses, internal or external, be readily available for replacement in the country of use. For Europe this means an IEC type. For smaller fuse types (i.e., 5 × 20 mm) there are fuseholders that can hold either U.S. or IEC fuses in the same fuseholder. For larger fuses it may be difficult to source one fuseholder to meet both the U.S. and IEC fuse sizes. Hence, circuit breakers are the best option in this case.

In addition to fuses or circuit breakers, thermal cutout and temperature-limiting devices are often required to limit temperature rise during fault tests. Temperature limiters or cutouts are used where an excess current draw is not sufficient to open a fuse, thereby causing an overheat condition. Transformers are regarded as temperature sensitive and may be subject to dangerous overheating due to a fault during a secondary short or overload. Motors may also overheat during a locked rotor or running overload test. Transformers and motors are often not protected by the primary fuse and overheat with a fault in the secondary. Fuses in the secondary or built-in thermal devices or both may be required to limit the hazards as a result of faults. Some standards require the use of temperature limiters in any case. These devices are considered critical to safety. Always specify a properly rated device that is suitable for its application and make sure they are European type-approved to ensure they meet all applicable standards for safety and reliability.

14. Stability and Mechanical Hazards

Under conditions of normal use, products shall remain "physically stable" to the degree that they cannot present a hazard to operators or service personnel. The various tilt and tip tests are not applicable for equipment that is intended to be secured to the building before operation, as specified in the installation instructions. When stabilizing means are provided to improve stability, when doors or drawers are opened, they must be automatic. If not automatic, suitable and conspicuous markings shall be provided to caution service personnel. The physical requirements depend on the applicable product standard. Many standards are available for machine guarding and protection. ITE standard EN 60950 gives a good description of the enclosure tests for stability and mechanical hazards and is the basis for this section.

Some examples of stability tests are:

- A unit shall not overbalance when tilted to an angle of 10° from its normal upright position (the doors and drawers are closed during the test).
- Floor-standing units with a mass of over 25 kg shall not tip over when a force of up to 20% of its weight, but not more than 250 N, is applied in any direction except upward, not exceeding 2 m from the floor.
- Floor-standing units shall not overbalance when an 800 N constant downward force is applied at the maximum moment to any horizontal working surface, or surface for obvious foothold, at a height not exceeding 1 m from the floor (the doors and drawers are closed during the test).

"Hazardous moving parts" shall be so arranged or guarded as to provide adequate protection against personal injury. Protection of the operator is most important, and suitable construction methods shall be provided to prevent access to hazardous parts. Permitted methods include locating the moving parts in areas that are not operator-accessible areas or enclosing the moving parts within an enclosure with mechanical or electrical interlocks that remove the hazard when access is gained. Service personnel shall be protected from unintentional contact with hazardous moving parts (and high voltage) during servicing of other parts of the equipment. It must not be possible to touch moving parts with the jointed test finger. In addition, openings preventing the test fingers entry shall be tested by a 30 N force from a straight unjointed version of the test finger.

When it is not possible to make hazardous moving parts directly involved in the process completely inaccessible and where the associated hazard is obvious to the operator, a warning may be considered adequate protection under certain conditions (see next section Warnings). Where the possibility exists that fingers, jewelry, clothing, hair, and so on may be drawn into the moving parts a method must be provided to stop the moving part and shall be placed in a prominent visible position and accessible where the risk is highest.

The edges and corners of enclosures shall be rounded or smoothed to prevent cutting hazards to operators. Furthermore, enclosures, handles, knobs, and the like shall have adequate strength to withstand rough handling during normal use and pass the relevant tests as described in the standards. Depending on the product and possible hazards the tests may include force tests, steel ball impact test, and drop tests (refer to standards).

Parts shall be adequately secured so that should any wire, washer, spring, screw, nut, or similar part fall out of position or come loose, it cannot reduce the distances over the reinforced or supplementary insulation levels specified in the standards. As mentioned, parts fixed in place by screws or nuts with self-locking washers or other means are not liable to come loose, and soldered wires are not considered adequately fixed unless they are held in place near the termination independently of the soldered connection.

Openings in enclosures tops, sides, and bottoms shall comply with the dimensional requirements of the relevant product or machine standards. The restrictions are necessary to prevent objects from entering the product via top or sides and prevent in the event of fire, flaming particles from exiting the bottom of the enclosure. Examples of enclosure opening sizes are:

- Top and side openings: (1) not exceed 5 mm in any dimension, or (2) not exceed 1 mm in width regardless of length, or (3) for tops be so constructed that vertical entry of falling object is prevented from reaching bare parts by means of a trap or restriction; or for sides, provided with louvers to deflect an external vertically falling object or so located that an object entering the enclosure is unlikely to fall on bare parts at hazardous voltages.
- Bottom openings: (1) baffle plate construction (see EN 60950), or (2) metal wire mesh not greater than 2×2 mm and a wire diameter of not less than 0.45 mm, or (3) size and spacing of holes in metal bottoms (i.e., 2-mm-diameter hole max. $\times$ 2 mm spacing min. for a 1-mm min. thick metal bottom).

15. Warnings, Instructions, and Languages

With the advent of strict liability in Europe the "duty-to-warn" viewpoint has recently been expanded in the courts to a considerable extent. If it is necessary to avoid hazards when operating, installing, maintaining, transporting, or storing equipment, safety information shall be provided by the manufacturer in the form of "warning symbols" on the product, along with "instructions" in the product documentation. Maintenance instructions are normally made available only to service personnel. For pluggable equipment intended for user installation, operating and installation instructions shall be made available to the user. Some of the common warnings are:

- **CAUTION**. Double-pole/neutral fusing. Disconnect power before servicing.
- **CAUTION**. This unit has two power cords. Remove both cords to disconnect power.
- **CAUTION**. High leakage current! Earth connection essential before connecting supply. (for stationary equipment)
- **CAUTION**. See installation instructions before connecting to the supply. (for equipment intended for connection to multiple-rated voltages or frequencies, unless means of adjustment is simple)
- **CAUTION**. Danger of explosion if battery is incorrectly replaced. Replace only with the same or equivalent type recommended by the manufacturer. Dispose of used batteries according to the manufacturer's instructions. (if product is provided with a replaceable lithium battery)
- **CAUTION**. Plug on power cord is disconnect device. The wall's socket-outlet shall be installed near the product and easily accessible. (for pluggable equipment without disconnect device)
- **CAUTION**. A readily accessible disconnect device shall be incorporated in the fixed wiring. (for permanently connected equipment without disconnect device)

The wording of electrical products instructions and products markings related to safety shall be in a language that is acceptable in the country in which the equipment is to be installed (ref; EN 60950, EN 61010-1, 73/23/EEC, 89/392/EEC, others).

Warning symbols must discourage operator access to compartments containing hazards and must warn service personnel of potential hazards when the hazard is not evident. Safety warnings shall be unequivocal by color, shape, and size and located as close to the hazard as possible. The warning symbols' size and colors are described in the relevant product safety and other supportive standards (ref; EN 60417, EN 60204-1, IEC 1310-1/-2, ISO 3461-1/3864/4196/7000, others). A triangle and pictogram distinguishes warning symbols from others that do not pertain to safety. The preferred method is to use black and yellow warning symbols, without text or signal words, so that the safety warning is unambiguous and negates the need to translate text. Other colors may be acceptable depending on the product-specific standard. If signal words and text are used, they should be translated into the appropriate languages.

When hazard symbols are used on the product, their meaning must be clearly explained in the documentation (i.e., user instructions), or alternatively such information shall be marked on the equipment.

Some of the more common warning symbols are:[1]

 CAUTION. Risk of electric shock

 CAUTION. Hot surface

 CAUTION. General warning (refer to accompanying documents)

Note. Except for General warning, signal words and text are not required when symbols are present (signal words are **CAUTION, DANGER, WARNING**). Depending on the applicable product standard, colors are typically black symbols and outlines (triangle) on a yellow background. The symbol shall be placed as close as possible to the hazard to warn the operator and/or service person before the hazard is accessed.

16. Flammability of Materials

Flammability requirements are intended to minimize the risk of ignition and the spread of flame, both within the product and to the outside. A "fire enclosure" is defined as a part of the product intended to minimize the spread of fire or flames

---

[1.] Symbols courtesy of Hazard Communication Systems, Inc. (HCS), Milford, PA.

within. A "decorative part" is a part outside the enclosure that has no safety function. Fire enclosures and components and parts inside an enclosure must be so constructed and make use of materials as to minimize the propagation of fire. Some materials and components are exempt from the requirements, such as small components, cable ties, and small parts mounted on PCBs. Other than the exempt items, all materials and components shall pass a flammability test or have a flammability rating, such as Class V-0 or 5V, V-1, V-2, HB or for foamed materials, HF-1, HF-2, HBF. Class V-0 and 5V are equivalent and the best ratings. V-0 and 5V are self-extinguishing within five seconds, and flaming drops or particles do not ignite surgical cotton. V-1 may take up to 25 seconds to extinguish and flaming particles may not ignite the cotton, so V-1 is a lower rating than V-0 or 5V. Class V-2 self-extinguishes within 25 seconds, however, the flaming particles may ignite the cotton.

Some of the common plastic materials and their minimum flammability classifications are:

- *Class V-1.* Fire enclosures for movable equipment not exceeding 18 kg
- *Class 5V.* Fire enclosures of movable and stationary equipment exceeding 18 kg
- *Class V-2.* Printed circuit boards (V-1 preferred, see Exempt)
- *Class V-2.* Wiring harnesses
- *Class V-2/HF-2.* Air filter assemblies (possible exceptions)
- *Class HB.* Decorative parts, mechanical and electrical enclosures, and parts of such enclosures if located external to fire enclosures
- *Exempt.* Components meeting flammability requirements of relevant IEC component standard, certain wire insulation (PVC/TFE/PTFE/FEP/Neoprene), small parts mounted on V-1 material (I/Cs, transistors, optocouplers, capacitors, etc.), one or more layers of insulation (adhesive tape, etc.), individual clamps, heat-resistant glass, nameplates, mounting feet, keycaps, knobs, lacing tape, twine, cable ties, ceramic materials, metals, others.

Manufacturers should select components and materials that have been previously flammability tested. Because of the popularity and focus on flammability in the United States (hot "flaming" oil test and "fire" enclosures), many materials and components have been tested and rated by UL. European testing and certification bodies generally recognize the UL flammability test results and classifications.

17. Electrical Safety Testing

Once the product meets all of the applicable design, component, and construction requirements detailed in the previous sections and the relevant safety standards, the electrical safety testing begins. For machinery, the subsequent sections and the relevant machine safety standards should also be met prior to testing.

📖   This section is only a brief overview of electrical safety tests. Refer to the standards for performing the tests and the pass/fail criteria.

Type tests are performed on a representative test sample or prototype of the equipment in question. The terms *type* and *model* are interchangeable. When a product complies with the component and construction requirements and successful test results are achieved, then all models of a given type are assumed to comply (e.g., of the same design and identification number [model number]). The equipment should be tested at normal operating conditions and carried out under the most unfavorable combination of the manufacturer's parameters for supply voltage, frequency, and so on and under full load (per standards; i.e., ±10% rated voltage).

The number of tests will vary from only a few tests, such as for equipment that uses all EU type-approved critical and safety components, to numerous testing for complex equipment using custom or nonapproved components and subassemblies. Using nonapproved components, where conformity is not verified, may require considerably more testing on the end-product manufacturer's part.

Some of the more common electrical safety tests for "products" include, but are not limited to:

- *Power consumption.* Establishes the total power (current) consumed by the product for input rating, circuit protection, and testing.
- *Ground continuity.* Tests the ability of the grounding system to withstand a high current (i.e., ≤ 25 A for 10 sec) between touchable metal covers/enclosures/doors and the mains ground (PE).
- *Voltage withstand.* The ability of a product's insulation to withstand high voltage between circuits (i.e., 3,000 Vac prim-to-sec for 1 min). This test is sometimes referred to as *hi-pot* or *electric strength test.*
- *Insulation resistance.* The system's insulation resistance between power circuits and the protective earth circuit (500 Vdc applied).
- *Earth leakage.* Measures any current that leaks onto the equipment enclosure and to a person or ground (3.5 mA maximum for movable products).
- *Capacitor discharge.* Measures the stored charge (residual voltages) on capacitors after a specified time (1 or 10 sec) to determine electric shock potential after power disconnection.
- *Temperature rise.* Used to isolate any components, plastic parts, and touchable surfaces that may attain excessive temperatures or exceed the allowable limits.
- *Stability and impact.* Tilt tests (10°) and side/top force tests are performed to ascertain the stability of equipment. Impact and/or drop tests determine the mechanical strength of products.
- *Abnormal operation and fault conditions.* Tests to limit risk of fire and electric shock from the equipment. Various abnormal operation and fault conditions are applied, such as electrical shorts/opens, component failures, mechanical faults, blocked vents, and overloads. The equipment's temperatures and electrical outputs are monitored during these tests. The product

must remain safe during and after the tests but is not required to still be in working order after testing.

- *Creepage and clearance.* Measurement of the minimum allowed distances between various circuits, such as primary-to-secondary, primary-to-ground, and within primary. Creepage and clearance distances between various circuits or parts, such as PCB traces, components, transformer windings, and wiring shall not be less than the minimum allowable dimensions when measured and with force tests applied.
- *Other tests as necessary.* Depending on the equipment's design and components used, numerous other tests may be required by the equipment safety standards.

The typical electrical safety tests for "machinery" include, but are not limited to (see also relevant machine safety standard[s]):

- Power consumption
- Ground continuity
- Voltage withstand
- Insulation resistance
- Capacitor discharge
- Functional
- Safety circuits, components, and safeguarding
- Other tests as necessary (temperature, leakage, noise, EMC, etc.)

The list assumes that all critical and safety components comply with their individual component safety standards and are acceptable for use in the end equipment. Additional tests may be necessary depending on the complexity of the equipment or components (i.e., specials, lasers, UV/microwave radiation, noise) and the environment where the equipment will be used (residential, industrial, hazardous locations). For test conditions and pass/fail criteria and other tests, refer to the relevant product/machine safety standard(s).

18. Production Tests

Routine safety tests are tests to which each individual product is subjected during or at the end of the manufacturing process to detect manufacturing variations and unacceptable tolerances in production and materials that could impair safety. These tests are performed to check the insulation between the primary circuits and accessible conductive parts and measure the resistance of the protective earth circuit. There are two production safety tests required for most products:

1. Electric Strength Test
2. Ground Continuity Test

The *electric strength test* (hi-pot) consists of applying to the equipment a sinusoidal AC voltage or an equivalent DC voltage (i.e., 1,500 Vac or 2,121 Vdc) between the primary and ground for 1 to 2 seconds, depending on the product standard. The tolerance of the voltage shall be +100 V, –0 V. No insulation breakdown shall occur during the test. An insulation breakdown is defined as any increase from the steady-state current measured during the test.

The *ground continuity test* is carried out by circulating a test current of not more than 25 A (typical for machines) or 1.5 times the current capacity of the product (i.e., 1.5 times the fuse/breaker rating), for the time required to obtain a meaningful reading through the parts to be tested and the ground (PE) terminal (e.g., between any user-touchable metal parts and the earth ground). The power supply cord shall not be included in the measurement.

All test results shall be kept available with the choice of support and format left to the manufacturer: separate forms or lists of equipment or grouped according to the most suitable parameters are equally acceptable. For all products manufactured and tested, the following data shall be filed:

- Date of test
- Model number
- Serial number or another identifier permitting unambiguous identification
- Value of voltage applied during the electric strength test[2]
- Value of the earthing circuit resistance and corresponding current value[2]
- Reference information that complete set of tests has/has not been successful

Depending on the equipment's design, additional tests may be required. Besides the applicable equipment safety standards, special Routine Test Standards exist, such as EN 50116 for Information Technology Equipment (ITE), EN 50106 for Household Appliances, EN 50144-1 for Handheld Motor Operated Tools, ENEC 303 for Luminaires, and CCA-201 for Certified products.

19. Additional Requirements for Machinery

This section addresses additional requirements according to EN 60204-1 (IEC 204-1) for the electrical safety of industrial machines. The requirements presented in the previous sections are applicable to machines and are for the most part contained in EN 60204-1 or other associated standards. I will point out some of the key electrical safety items that are often overlooked by the beginning machine safety inspector. EN 60204-1 does not cover all machine safety requirements for guarding, interlocks, control, and so on that are also applicable and listed in the Machinery Directive.

Notice. To ensure the machine's conformity with all the electrical safety requirements, a complete assessment according to EN 60204-1 and other applicable standards must be performed by qualified safety persons. Additional requirements may apply depending on the machines use, environment, and machine safety standards.

---

[2] Alternatives to the values are permitted if the pass/fail criteria is described elsewhere in the test record.

The essential factors relating to electrical aspects of machines are to promote: (1) safety of persons, property, and machine; (2) consistency of control response; and (3) ease of maintenance.

✓ *Safety of the operator and service personnel is most important.* High performance should not be obtained at the expense of these essential factors.

A. Protective Measures *Protective measures* shall be employed to effectively:

1. Protect the "operator" from shock.
2. Protect the "equipment" from overload conditions.

The equipment design and overall enclosure shall provide protection of persons against electric shock from:

- *Direct contact.* All measures for protection of personnel that may arise from direct contact with live parts of the electrical equipment.
- *Indirect contact.* Protection of personnel from hazards that may arise in the event of an insulation failure (single fault) between live parts and exposed conductive parts.

Protection against direct contact with live parts is fulfilled in a number of ways, such as enclosures, guards, interlocks, insulation, and the use of SELV or PELV circuits.

The following measures shall be considered to protect the equipment from:

- Overcurrent arising from a short circuit
- Overload currents
- Abnormal temperature
- Loss or reduction in supply voltage
- Overspeed of machine elements

The equipment and personnel protection measures and device types, usage, arrangement, and operation of devices are described in EN 60204-1 and other related standards.

All fuses used in machines, internal and external, must be of an European IEC type. U.S. fuses are *not* readily available throughout Europe, despite what we hear. Circuit breakers may be the best option.

B. Mains Disconnect Switches *Mains disconnect switches* are required on all machines to isolate the entire electrical equipment. The OFF position must be lockable, IEC symbols for ON/OFF positions must be present and meet the relevant EU standards for construction and operation, which is assured if the device is EU type–approved as a mains disconnect or as a combined E-stop/disconnect device (rotary lockable type, red/yellow, etc.). Star-delta, reversing, multi-pole, and U.S. knife-type (pull-down) switches are not acceptable. The switch must be capable of switching the stalled current of the largest motor and at the same time total currents of remaining loads.

C. Emergency Stop Switches   *Emergency stop (E-stop) switches* shall be provided on every machine to avoid injury to personnel and the surroundings or damage to the machine. The E-stop shall stop the dangerous elements of the equipment or stop the entire machine as quickly as practicable. The purpose of the E-stop is to stop the machine in case of danger and disconnect it from the supply voltages. When operating, all loads that may lead to a hazard to personnel or damage to the machine must be disconnected by deenergization so as to deenergize contactors, relays, or the undervoltage release of the main disconnect switch.

💣   *Emergency stop devices are required for machinery!* They shall have a red button on yellow background, be self-latching, positively opening, readily accessible to operator, manually activated, require manual reset. Resetting the E-stop switch shall not restart any part of the machine.

There are several possible E-stops. The palm, or "mushroom head" type is the most popular. The actuator must be readily visible and easily reached by the operator from the working and operating positions. Several E-stop devices may be required to cover all the machines' working or operating positions. The mushroom head or other actuator must be red and the background yellow to clearly identify the device as an emergency stop switch. A combined supply disconnect/E-stop device also exists and must meet all the E-stop and color requirements. The requirements for E-stops are defined in various standards, such as EN 60204-1 and EN 418.

D. Fault-Tolerant Components and Safety Circuits   *Fault-tolerant components* and properly designed *safety circuits* are necessary to ensure safe and reliable machine control (refer to EN 954, 364, 292, 1050, 60204-1, etc.). Common terms associated with these special machine components are *control reliable, fault tolerant,* and *fail-safe,* which means that if a failure occurs (single fault) it will always fail to a safe condition (e.g., fail-safe). The machine's circuit design and components must ensure user protection from all safety hazards, such as moving parts or high voltage. The requirements for these highly critical components go beyond the requirements for the typical critical components, both of which must be met for machinery. All components within a safety circuit are of special concern, and it's these components, such as electromechanical relays, contactors, guard/door interlocks, and E-stops switches, that are invariably the weak point of the circuit. Components in this category have additional requirements that must be met for safe "control reliability." Refer to the proper standards for these special components, such as EN 348, 947, 50100, and 60204-1.

To illustrate some common misconceptions, a few examples of compliant and noncompliant industrial-type components are discussed in the next sections. Fault-tolerant components, that have been EU type–approved for proper classification, such as positive opening, guarded actuator, redundancy, cross-monitoring, or fault detection, are preferred and in some cases mandatory. Testing nonapproved components (CE is *not* an approval) to verify their conformity or nonconformity is the higher risk (of failure) alternative and usually costs considerably more time and money.

E. Transformers   *Safety and isolating transformers* are frequently used in electrical and electronic products and machines. Transformers are a common cause for noncompliance (e.g., if nonapproved). The major reasons for noncompliance are overheating and construction. Tests and construction requirements are detailed in the relevant equipment standards and/or transformer standards EN 60742 (IEC 742). Transformers that comply with EN 60742 usually are in compliance with the transformer requirements contained in the equipment standards. The standards have strict construction and testing requirements that are sometimes overlooked by transformer manufacturers or machine builders who source them. For example, the standards require that transformers must be protected against overheating during a short circuit and while under an overload condition. The standards limit the allowed temperature rise in transformers during normal conditions and in the event of an internal or external fault. Some common problems with nonapproved transformers are: displacement of windings, improper wire terminations, inadequate creepage/clearance prim-to-sec, loose parts bridging prim-to-sec, low-grade or improper class of insulation, thermal limiter/cutout, and secondary fusing missing. Protective devices are recommended, or mandatory in some cases, for effective protection against overloading and short circuits.

EN 60204-1 requires the use of a control transformer on machines with more than five electromagnetic coils. The transformer must be installed after the mains switch to ensure isolation. Small transformers are not typically suited for control circuits since they are designed for simple resistive loads. Overcurrent protection is required in accordance with standards EN 60742, 60204-1, IEC 76-5. The type and setting of the overcurrent device should also be in accordance with the recommendations of the transformer supplier. The preferred secondary voltages are 24/48/115/230 V, 50/60 Hz. Fault-free operation must be verified prior to using lower voltages.

F. Motors   Motors, like transformers, are another problem facing machine builders. Various standards specify construction and overload protection requirements for motors (EN 60034, 60742, 60664, 60204-1). Motors must meet the requirements for dimensions, construction, insulation class, connection, overload protection, ingress protection, markings, abnormal testing, and other parameters.

Effective motor protection is crucial and achieved by using overload protection devices, temperature-sensing devices, and current-limiting devices. The protection devices shall protect motors against:

- Excessive temperature rise
- Rapid destruction during start-up or in the locked-rotor condition
- Unacceptable reduction in service life
- Nuisance tripping during normal operation

Overload protection is recommended for all motors, especially coolant pump motors, and shall be provided for each motor rated at more than 0.5 kW. Continuously operating motors over 1 kW shall be protected against overloads and against a stalled rotor state with built-in thermal sensors typically employed (EN 60034-11). The use of appropriate protection devices for special-duty motors (e.g., rapid traverse, locking, rapid reversal) is recommended. For motors with ratings less than or equal to 2 kW, overload protection may not be required.

U.S. wire-nuts are not permitted for wire connections. Fixed terminals for wire termination, within a motor junction box, is the preferred method. If the motor is nonapproved, as is sometimes the case for larger motors, additional testing and review will be necessary to verify conformity of the motors construction and its protection devices.

Motors shall meet the construction requirements (spacings, materials, etc.), employ fixed terminals (no wire-nuts) for wire connections, utilize effective protection devices, and pass the relevant tests.

G. Wiring   Wiring conductors and cables shall run from terminal to terminal without splices (no wire nuts), and the terminations of multicore cables adequately supported will be to reduce undue strain on the terminations of the conductors. Internal wiring should be identified by the proper colors and numbers, and identification tags shall be legible and permanent. The conductors shall be identifiable at each termination and in accordance with the technical documentation. The minimum cross section of single or multicore wires is 0.75 mm^2 and flame-retardant wire insulation should be used (e.g., V-2 min). The color coding is listed in EN 60204-1 and IEC 757, with the following colors specified:

Neutral	Light blue
Safety and protective earth (PE)	Green-and-yellow
AC or DC power circuits	Black
AC control circuits	Red
DC control circuits	Blue
Interlock circuits (external source)	Orange

Note. Green or yellow should not be used since there is a possibility of confusion with bicolor green-and-yellow (PE).

All live power conductors, except the earthed neutral, shall employ overcurrent protection selected as low as possible but adequate for the anticipated start-up currents. Control circuits conductors connected directly to supply voltages and circuits feeding control circuit transformers shall be protected. Refer to EN 60204-1 for more information.

H. Protective Earth   The *protective earth (PE) terminal* for the external ground (earth) conductor shall be marked with the letters *PE*. The PE designation is restricted to the terminal for bonding of the external protective conductor (green-and-yellow) of the incoming supply. Other grounds may not be placed on the same ground terminal as the PE conductor. To avoid confusion, light blue is reserved for neutral. Green-and-yellow is for safety and PE ground only and shall not be used for low-voltage (< 50V) or EMC/RFI grounds.

The wiring "terminals" shall be appropriately and plainly marked and correspond with the circuit diagrams. Terminals shall be of a IEC type, which means touch-safe. All connections must be double fixed to secure them from accidental loosening, especially the protective earth conductor. Placing two or more wires into one terminal is permitted only in the cases where the terminal is designed and tested for that purpose. Terminal blocks shall be so mounted that wiring does not cross over the terminals. Selecting an EU type–approved terminal block is the best option to ensure conformity without further testing. Wire splicing is prohibited, and U.S. wire-nuts are not allowed as they do not meet the IEC requirements for wire connections. They are not considered as a reliable termination means (e.g., not adequate in securement, identifiable, fixed from moving).

💣   U.S.-style twist wire-nuts do not meet the EU requirements and shall not be used.

I. Access Areas   *Operator access area* is any area to which, under normal operating conditions, access can be gained without the aid of a tool (e.g., by a person's hand or fingers alone). Opening a hinged door or removing a cover by hand, without a tool, makes the area behind the door/panel an operator access area, and all hazards shall be adequately guarded or interlocked to remove the hazards before access. Fixing the door or cover in place with screws is one way of protecting persons, but if door or cover interlock switches are used to reduce a hazard (high voltage, moving parts, etc.), they must be of a fail-safe type and nonoverridable by the standard test finger. The nonsafety type that may be activated (ON/OFF) by a person's finger through a push action and/or pulled into a locked ON position is not permitted. In addition, the restoration of the interlock shall not initiate machine motion or operation if this could give rise to a hazardous condition.

✓   Machine builders sometimes forget that tops and bottoms may be necessary on the equipment to keep persons and foreign objects out, and to minimize the spread of fire from within.

J. Enclosures   Selecting an electrical control enclosure that meets all of the relevant requirements is one of the problems facing machine manufacturers. Enclosures must be mounted to facilitate accessibility and maintenance and protected against external influences under normal use conditions. This can be accomplished only if the enclosure meets the EU standards and is used properly in the equipment. EN 60204-1, EN 60529, and other standards are very specific concerning the enclosures' location, construction, component types and placement, and grounding and ingress protection (IP). For most machines, enclosures must provide a minimum protection of IP 54, which means protection against dust and splashing water (Figures 6-6 and 6-7). The first digit refers to the dust rating and the second to the water. If an X is signified in the rating it means that the test was not performed (i.e., IP X4 says it was water tested only). IP 54 is the minimum degree of protection for most machine enclosures with a higher or lower degree allowed depending on the installation conditions or environment. In addition to the IP rating, other construction requirements exist. All openings, including those in the bottom or those for mounting purposes, shall be closed in a manner ensuring the specified degree of ingress protection. Components and devices may not be placed on doors except those for operating, indicating. Don't forget that the door must have a ground wire or strap since the hinges are not considered adequate for grounding purposes. Enclosure doors should have vertical hinges (lift-off type) and captive fasteners. In general, if the enclosure bears an EU type–approval mark it may be considered to comply with the IP and construction requirements without further testing. Also, enclosure manufacturers whose product lines are EU type-approved usually offer complimentary products with EU approvals, such as air conditioners, lights, power strips, door interlocks, and grounding kits.

> *Caution!* U.S. NEMA class and European IP ratings are not always equivalent (NEMA 12 ≠ IP 54)! Both the dust and water tests must be performed according to EN 60529 with passing results. Enclosures often fail the IP requirements, especially the *dust* tests. It is best to use a type-approved and marked (VDE/TUV) enclosure to ensure conformity.

K. Functional Markings   *Functional markings* should be used for all controls, switches, and indicators whenever possible, unless their function is obvious. Proper markings and colors are essential if the control or indicator in question is related to safety. Red, green, and yellow are important push-button and indicator colors and strictly controlled by the standards for safe operation of the equipment. A red light is used to indicate an *emergency* condition and instructs the user to operate the emergency-stop (E-stop), thereby disconnecting power to the equipment. A green light shows a normal or run condition and is commonly used as a power ON indicator. Yellow is used as a *warning* to indicate an impending hazardous condition may occur and instructs the user to take appropriate action. Most safety standards

1st digit	Protection from solid objects	2nd digit	Protection from moisture
0	Non protected	0	Non protected
1	Protected against solid objects greater than 50mm	1	Protected against dripping water
2	Protected against solid objects greater than 12mm	2	Protected against dripping water when tilted up to 15°
3	Protected against solid objects greater than 2.5mm	3	Protected against spraying water
4	Protected against solid objects greater than 1.0mm	4	Protected against splashing water
5	Dust protected	5	Protected against water jets
6	Dust tight	6	Protected against heavy seas
–		7	Protected against the effects of immersion
–		8	Protected against submersion

Notes:

EN 60529 outlines an international classification for the sealing effectiveness of enclosures of electrical equipment against the intrusion into the equipment of foreign bodies (i.e., dust, tools, fingers) and moisture. This classification system utilizes the letters "IP" (Ingress Protection) followed by two digits. An "X" is used for one of the digits if there is only one class of protection; i.e. IP X4 which addresses moisture resistance only.

EN 60529 does not specify sealing effectiveness against the following: the risk of explosions; certain types of moisture conditions, e.g., those that are produced by condensation; corrosive vapors; fungus; vermin.

FIGURE 6-6 Ingress Protection (IP) Levels per EN 60529 (IEC 529). *Courtesy of Panel Components Corp., Oskaloosa, IA.*

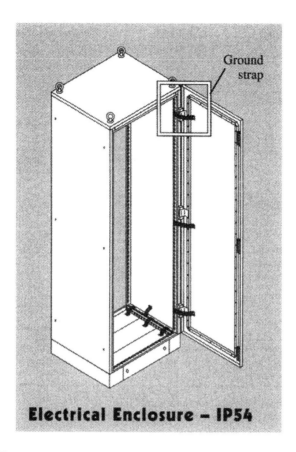

FIGURE 6-7 Electrical Enclosure–IP54. *Courtesy of Rittal Corporation, Springfield, Ohio.*

will cover specific information required for safety functions and markings for the product in question with reference to other standards for details. Figures 6-8 and 6-9 show push-button and indicator light colors and their meanings.

Markings for machinery are the best defined, and various standards detail the accepted conventions (e.g., EN 60073 and EN 60204-1 for colors and meanings, IEC 1310-1/-2 and ISO 7000 for actuator movements, hazard symbol types and sizes).

L. Item Designations *Item designation* markings are required within the equipment for machine parts and terminals to ensure ease of identification and maintenance. These markings help service personnel to correlate the parts in the equipment to the different diagrams, parts lists, circuit descriptions, and instructions. All items, such as basic parts, components, terminals, subassemblies, and

Color	Meaning	Explanation	Examples and application
Red	Emergency	Actuate in case of hazardous condition or emergency	Emergency conditions Initiation of emergency function Used for E-Stop actuators
Yellow	Abnormal	Actuate in case of abnormal condition	Intervention to suppress abnormal condition Intervention to restart an interrupted automatic cycle
Green	Safe	Actuate in case of safe situation or to prepare normal	Reserved for functions indicating a safe or normal condition
Blue	Mandatory	Actuate in case of condition requiring mandatory action	Reset function
White	No specific meaning assigned	For general initiation of functions except emergency stop (see also note)	Start/On (preferred) Stop/Off
Grey			Start/On Stop/Off
Black			Start/On Stop/Off (preferred)

Note: Where a supplemental means of coding (e.g., texture, shape, position) is used for the indentification of push-button actuators, then the same color White, Grey, or Black may be used for various functions (e.g., White for Start/On and Stop/Off actuators).

FIGURE 6-8 Push-Button Colors and Their Meanings

assemblies shall be plainly identified with the same symbols as shown in the technical documentation (Figure 6-10). The item designations are generally assigned when the circuit diagram is drawn. The designations are located adjacent to the items and on or adjacent to terminals (not on replaceable components). The item designation markings (labels, etc.) shall be permanent and visible after the components and wiring are in place. The item letter codes and methods for establishing these designations are recommended in IEC 750 and EN 61346-1/-2.

An *item designation* is a distinctive code that serves to identify an item in a diagram, list, chart, and on the equipment. A *basic part* is one piece (or several pieces joined together) that cannot normally be disassembled without destroying its function, such as an integrated circuit or resistor. A *subassembly* is two or more basic parts that form a portion of an assembly, replaceable as a whole, but having a part or parts that are individually replaceable, such as overcurrent protective device, filter unit, terminal board. An *assembly* is a number of basic parts or subassemblies or any combination thereof joined together to perform a specific function, such as electrical generator, audio amplifier, power supply, or switchgear assembly. *Items* are defined as a basic part, component, equipment, or functional unit and are usually denoted by graphical symbols on diagrams. Resistors, relays, generators, amplifiers,

Color	Meaning	Explanation	Action by operator	Examples of application
Red	Emergency	Hazardous condition	Immediate action to deal with hazardous condition (e.g., by operating E-Stop device)	Pressure/temperature out of safe limits Voltage drop Breakdown Overtravel of a stop position
Yellow	Abnormal	Abnormal condition Impending critical condition	Monitoring and/or intervention (e.g., by reestablishing the intended function)	Pressure/temperature exceeding normal limits Tripping of protective device
Green	Normal	Normal condition	Optional	Pressure/temperature within normal limits Authorization to proceed
Blue	Mandatory	Indication of condition which requires action by the operator	Mandatory action	Instruction to enter preselected values
White	Neutral	Other conditions; may be used whenever doubt exists about the application of Red, Yellow, Green, Blue	Monitoring	General information

FIGURE 6-9 Indicator Light Colors and Their Meanings

power supply units, and switchgear assemblies may all be described as items for designation purposes. *Terminal designations* are applied to the conducting parts of an apparatus (screw terminal, terminal block, quick-connect tab) provided for electrical connection to the circuits and conductors. *Aspect* is the specific way of selecting information on or describing a system or an object of a system, such as:

- What the system or object is doing (function viewpoint);
- How the system or object is constructed (product viewpoint); or
- Where the system or object is located (location viewpoint).

The item designations and symbols may be placed in a *block format* with each block consisting of latin letters or arabic figures or both, uppercase letters being preferred. The blocks may be preceded by a prefix sign. Prefix signs are used to distinguish the various designation blocks and enable the blocks to be combined in any suitable manner:

Prefix Sign	Location	Type/Number/Function	Example
=	Block 1	Higher level (function)	**=T2**
+	Block 2	Location	**+D126**
-	Block 3	Item (aspect)	**–K5**
:	Block 4	Terminal	**:14**

Letter Code	Kind of Item	Examples
A	Assemblies, subassemblies	Amplifier using discrete components, magnetic amplifier, laser, maser, printed wiring board
B	Transducers, from nonelectrical to electrical quantity or vice versa	Thermoelectric sensor, thermo cell, photoelectric cell, dynamometer, crystal transducer, microphone, pick-up
C	Capacitors	
D	Binary elements, delay devices, storage devices	Digital I/C's and devices, delay line, bistable element, monostable element, core storage, register, magnetic tape recorder, disk recorder
E	Miscellaneous	Lighting device, heating device, device not specified elsewhere in this table
F	Protective devices	Fuse, overvoltage discharge device, arrester
G	Generators, power supplies	Rotating generator, rotating frequency converter, battery, oscillator, quartz-oscillator
H	Signaling devices	Optical indicator, acoustical indicator
J	-	-
K	Relays, contactors	
L	Inductors, reactors	Induction coil, line trap, reactors (shunt and series)
M	Motors	
N	Analogue elements	Operational amplifier, hybrid analogue/digital device
P	Measuring equipment, testing equipment	Indicating, recording, and integrating measuring devices; signal generator; clock
Q	Switching devices for power circuits	Circuit-breaker, disconnector (isolator)
R	Resistors	Adjustable resistor, potentiometer, rheostat, shunt, thermistor
S	Switching device for control circuits, selectors	Control switch, push-button, limit switch, selector switch, dial contact, connecting stage
T	Transformers	Voltage transformer, current transformer
U	Modulators, changers	Discriminator, demodulator, frequency changer, coder, inverter, converter, telegraph translator
V	Tubes, semiconductors	Electronic tube, gas-discharge tube, diode, transistor, thyristor
W	Transmission paths, waveguides, aerials	Conductor, cable, busbar, waveguide, waveguide directional coupler, dipole, parabolic aerial
X	Terminals, plugs, sockets	Connecting plug and socket clip, test jack, terminal board, soldering terminal strip, link, cable sealing end and joint
Y	Electrically operated mechanical devices	Brake, clutch, pneumatic valve
Z	Terminations, hybrid transformers, filters, equalizers, limiters	Cable balancing network, compandor, crystal filter, network

Notes

In the General Index of IEC 817-1: Graphical Symbols for Diagrams, Part 1: General Information, General Index. Cross-reference Table (1985), the commonly used letter codes are given for items with standardized graphical symbols.

If more than one designation is possible, because an item can be described by more than one name, the more specific designation should be used.

FIGURE 6-10 Letter Codes for Item Designations

For complex items or installations a block format is utilized, such as **–K3M** for relay K3 used for monitoring function or **–Q2Q1** for circuit breaker Q2 with main contact assembly Q1. The following sequence is preferred:

$$= \boxed{1} + \boxed{2} - \boxed{3A\ 3B\ 3C} : \boxed{4}$$

If no confusion can arise, the intermediate prefix sign(s) may be omitted, and in many cases a simplified marking (block 3) is sufficient. Examples of simplified markings are **K3** for contactor K3, **M4** for motor M4, **Q6** for circuit breaker Q6, and **K5M** for relay K5 with a main function. The following combinations are permitted:

$$\boxed{3B} \ \text{or} \ \boxed{3A\ \ 3B} \ \text{or} \ \boxed{3B\ \ 3C} \ \text{or} \ \boxed{3A\ \ 3B\ \ 3C}$$

M. Warning Symbols   *Warnings* are required for all hazards to notify personnel (the "duty-to-warn") such as on any enclosure panel or door that does not clearly show that it contains a hazard. A black triangle and pictogram on a yellow background, in accordance with the standards, shall be used. The service persons must also be warned of any possible hazards before they access a compartment, such as high voltage or energy, moving parts, high temperature, or laser radiation. Additional warning symbols may be required on internal covers and adjacent to hazards within the compartments to protect against accidental or inadvertent contact. See also Section 15, Warnings.

In some cases, a warning may be considered adequate if it is not possible to make hazardous moving parts directly involved in the process completely inaccessible and where the associated hazard is obvious to the operator. In such a case, where fingers, jewelry, clothing, etc. can be drawn into the moving parts, a warning shall be provided in a visible and prominent position for stopping the moving part. An example of where it may not be possible to guard the hazard is the visible moving parts of a paper cutter where hand-feeding is required.

User protection is paramount! Warnings are only permitted when no other means are possible and may not take the place of a safe design. For example, a **MOVING PARTS WARNING** is allowed only when the hazard is directly involved in the production process and there are no other possible options (i.e., guards, interlocks, stop-switch, sensors). Therefore, if a guard or other protection means is possible, it must be employed.

N. Manuals   Machine manufacturers shall deliver a machine that is safe to operate and, therefore, the operators instructions shall be in the language of the country the machine is intended for. For machinery, the complete operators manual, with safety instructions, shall be in a language acceptable in the country for its operation (ref; EN 292-2, EN 60204-1, 89/392/EEC, others). Translation of the installation and maintenance instructions may also be necessary depending on the language of the installer and service personnel. All manuals requiring a country-specific language must be translated by the equipment manufacturer/supplier and made available prior to putting the product or machine into service.

The equipment manufacturer or supplier is responsible for translations. Translation of the operators manual is a must! Translation of service and installation manuals may also be required if requested by the customer. The language and translation requirements are mandatory and the equipment buyer or user may not accept liability for language translations.

O. Technical Documentation   *Technical documentation* necessary for the installation, operation, and maintenance of the machine shall be supplied. The information must be provided by the machine supplier to the buyer before delivery of the equipment. Circuit documentation (diagrams, tables, descriptions) serve to explain the function of circuits, power connections, and process-oriented functions (see IEC 204-2, 617-1, 848, 1082).

Documentation shall include:

1. Description of machine, installation and mounting, and the connection to the electrical supply;
2. Electrical supply requirements;
3. Physical environmental information (i.e., lighting, vibration, noise levels, atmospheric contaminants) where appropriate;
4. Circuit diagram(s);
5. System or block diagram(s) where appropriate;
6. Information (where appropriate) on programming, sequence of operation and inspection, frequency and method of functional testing, guidance on adjustments and maintenance (especially on the protective devices and circuits), parts list, in particular the spare parts;
7. Description of the safeguards, interacting functions, and interlocking of guards for hazardous movements; and
8. Description of safeguarding means and methods where the safeguards are suspended.

Refer to the appropriate standards for the documentation requirements.

This section is an introduction to key machinery requirements according to EN 60204-1 and other related standards. Other requirements and standards may apply, including, but not limited to, risk assessment, safety circuits and components, guarding, electrical, mechanical, radiation, chemicals, gases, documentation, and testing.

Safety Checklist

The Safety Checklist	Company : _____ Factory : _____ Eng : _____ Date : _____	Product : _____ _____ Model No : _____ Serial No : _____	Directive(s) : _____ Standard(s) : _____ _____ _____	
No	**Clause**	**Requirements**	**Comments**	**✓**
1		Critical components comply with EN/IEC component standards and verified by: Approval marks, certificates, other* (See 2 and 3)		
2		Components conformity verification: Approval marks visible on components Certificates and reports available (upon request) Restrictions for use complied with, per certificate Components needing additional IEC confirmation:* Transformer, motor, other; Checked by: submittal, testing, other		
3		Constructional Data Forms (CDF) complete: Manufacturer, part number, and rating verified Approval mark(s) shown on CDF (*not* CE or IEC) * Certificates on file (Attach Constructional Data Form[s])		
4		Equipment shock classification: Class I, Class II, Class III Equipment mobility classification: Hand-held, movable, stationary, fixed		

The Safety Checklist	Company : _____ Factory : _____ Eng : _____ Date : _____	Product : _____ _____ Model No : _____ Serial No : _____	Directive(s) : _____ Standard(s) : _____ _____ _____

No	Clause	Requirements	Comments	✓
5		Input power meas = _____ VAC, _____ A, ___Hz or, _____VDC, _____ mA, or for three phase; 3/PEN AC 400 V ___ A 50 Hz (four wire), or 3/N/PE AC 400 V ___ A 50 Hz (five wire), or other		
6		Rating label: Company, model number, voltage, amperage and frequency, IEC symbols, visible, durable, next to input, date code (machines), other		
7		Markings: On/Off symbols, fuses, outlets, colors of indicators lights and buttons, other Machinery: E-stop, function and controls, other		
8		Single-fault and abnormal tests: Short/open nonapproved critical components xfmr outputs, blocked vent, others (Test results attached)		
9		Wiring: V and A rating (size), UL insulation, fixed cord <HAR>, grn/yel safety, dbl fixed terms, 600 V for primary or 300 V w/sleeving Machinery: ident. by colors or numbers Power: inlet/module or fixed cord (<HAR>); IEC strain relief, IEC terminals (touchsafe), L, N, PE markings, double fixing, other		

The Safety Checklist	Company : _____ Factory : _____ Eng : _____ Date : _____	Product : _____ _____ Model No : _____ Serial No : _____	Directive(s) : _____ Standard(s) : _____ _____ _____	

No	Clause	Requirements	Comments	✓
10		PCB/transformer insulation: reinforced (prim-sec), basic (prim-grd), other Wire separation (prim-sec): distance or thickness (2 layers) or 3 layers, or other		
11		PE ground is isolated, connection secure, and PE ground symbol adjacent to stud: "⊕" PE symbol (one), "⊕" other grounds, metal encl grd'd grn/yel wire for safety grd's, other		
12		Power disconnect: Double pole switch, circ brkr, cord plug, other		
13		Mains circuit protection: Fuse, circuit breaker, type (T, F, etc.): ___ Size: ___ A ____Vac/DC Thermal protectors (motors, transformers): type: _____ size: _____ other info Fuse/therm protector approvals and locations: VDE/S or other, location(s)		
14		Stability and mechanical requirements: impact tests drop tests, hand-held products stability of nonfixed equipment enclosure openings access hazard guarding (refer to standards for requirements and tests)		

The Safety Checklist	Company : _____ Factory : _____ Eng : _____ Date : _____	Product : _____ _____ Model No : _____ Serial No : _____	Directive(s) : ____ Standard(s) : ____ _____ _____
No / **Clause**	**Requirements**	**Comments**	**✓**
15	Warnings: IEC symbols (black/yellow) leak- age, laser, dual fusing, high voltage/ energy, moving parts, high temp, other(s) Instructions: text in manual(s) and warning sym- bols on equipment Language(s): English, French, German, other(s) as required		
16	Plastics flammability: Enclosure, wire insulation, decora- tive parts, other parts Tested by: ____, cert's on file		
17	Electrical safety tests: Power consumption (see 5) Ground continuity Voltage withstand Insulation resistance Earth leakage Capacitor discharge Temperature rise Stability and impact (see 14) Creepage and clearance (see 10) Abnormal and fault conditions (see 8) Others (Test data attached)		
18	Production tests: Hi-pot (1,500Vac/2,120Vdc for 1 sec) Ground Continuity (25A or 1.5 max inp) Other		

The Safety Checklist	Company : _____ Factory : _____ Eng : _____ Date : _____	Product : _____ _____ Model No : _____ Serial No : _____	Directive(s) : _____ Standard(s) : _____ _____ _____
No | **Clause**	**Requirements**		**Comments** | ✓

No	Clause	Requirements	Comments	✓
19		Additional machine requirements, complies with: 15 items in "Safety Guide" (A–O) Machine safety standard; _____ EN 292-1/-2, 614-1, 954, 1050 EN 60204-1 Other applicable standards (Refer to electrical and machine safety standards; technical documentation and reports attached)		

Notes on using the Design Checklist:
- * CE marking and declarations of conformity should not be considered as evidence of compliance for components, products, or machines.
- ✓; P = Pass, F = Fail, N = Not Applicable, placed in right column.
- Identify clause(s) from the applicable standard(s) and place in second column. More than one standard or clause may apply.
- Comments should include how and why the product passed/failed (P/F) or explain why the clause is not applicable (N).
- Attach Constructional Data Form (CDF), test results, and other support documentation to the Safety Checklist.
- *Notice.* The Safety Checklist is for a preliminary safety review only and does not take the place of a complete assessment or test report(s). Other requirements may apply. Refer to appropriate standards and directives for final assessment and completion of test report(s).

© Lohbeck

CHAPTER 7

Conclusion

*We can forgive a child who is afraid of the dark; the real tragedy of life
is when adults are afraid of the light.*

PLATO (CA. 400 B.C.)

Focus on Standards

The goal of the *CE Marking Handbook: A Practical Approach to Global
Safety Certification* is to increase everyone's awareness of the real meaning of the
CE marking and the importance of producing safe products. The myths and realities
of European Conformity have been presented, and with this knowledge manufacturers, exporters, and consumers can make sound decisions.

If this is your first experience with European Conformity, CE, or even if you
have been working in the field for some time, understanding the rules may have
seemed like an insurmountable task. With all the misinformation and confusion surrounding the CE marking this attitude is certainly reasonable, but I hope you now
have a better understanding of the real meaning of the CE marking and the technical
rules for safe product design and the required assessment procedures that ensure
product conformity. Focus on *standards* to attain the presumption of conformity for
your products.

> Directives tell us *why* we must comply (consumer safety/EMC)
> and *what* may happen if we ignore the laws (withdraw products). But
> it's the European *standards* that show us *how* to comply (design and
> assessment).

CE's Credibility at Risk

Evidence to date leaves the credibility of the CE marking in doubt. Based
on my numerous safety/EMC inspections of equipment (> 75% fail) and seeing

the results of policing activities by enforcement authorities in Sweden (> 50% fail), Germany (> 60% fail), Finland (> 75% fail), and elsewhere, it is my view that the majority of products bearing "only a CE marking" *do not comply* with the European safety and EMC requirements or are questionable at best.

The art of oversell has reached new heights since the introduction of the so-called coveted CE marking. The CE marking is only the manufacturer's *self-declaration,* a do-it-yourself approach, *and does not, in and of itself, guarantee conformity* and may not meet the customer's expectations. CE markings do not ensure conformity and should not be used as a cover-up. Unfortunately, some manufacturers use the CE marking and declaration as a shield for what they did or did not do. The most infamous line is "The CE marking is all you need." In the legal sense this claim is true; in the technical sense the claim is unfounded. Legally, the CE marking allows for the distribution of products; practically speaking, far more is required to satisfy technical safety requirements and users concerns. Also, CE does *not* limit liability, and it is *not* intended for marketing, sales, or quality assurance purposes.

? *Question:* If three products are placed on the European market with "only the CE marking," what's the difference between the products?

☹ *Answer:* One of the products complies with Europe's safety and EMC requirements and the other two don't!

As a manufacturer or user of components, products, or machines, do not allow yourself to fall into the trap of unCErtainty. Consumers and inspectors have the right to demand positive evidence of conformity from independent accredited parties. CE markings and declarations alone are not adequate to verify the safety or EMC conformity of components or equipment.

📖 Always obtain positive evidence of conformity (EU approval marks, certificates, test reports). Play it safe and you won't get burned!

Self-Test: "Where Are You Now?"

You may be surprised and even discouraged at what first appears to be the work involved to achieve European Conformity for your products. At this point you can either fight it and continue to ask, "Why me?" or simply apply the rules and "accept them." Remember, for the most part the world is or soon will follow the European safety, EMC, quality, and environmental rules. It is sometimes difficult for us to change our ways of thinking, but once we learn and apply the principles to the design and manufacturing of the first product, achieving subsequent product conformity will be much easier—if not automatic.

Now that you have a greater appreciation of product assessment and testing for the CE marking, take a self-test just for fun (Figure 7-1). You may check any box below, but the sooner you attain acceptance, the sooner you can ship products to the European Union.

"Where are you now?" (☑ check one or more boxes):
☐   Denial and Isolation.
☐   Anger and Rage.
☐   Bargaining.
☐   Depression.
☐   All of the above!
☐   Acceptance (affix "CE" marking).

FIGURE 7-1   Self-Test Checklist

Total Harmonization and Product Certification

Meeting the European Conformity safety testing and design requirements for CE marking is as simple as one, two, three: (1) components, (2) construction, and (3) testing. A product's compliance with the European requirements relies on the application of good common sense, experience, and knowledge of the European safety philosophy and standards. Of utmost importance is "consumer protection" via conformity to the "harmonized standards." The standards and directives were not established to get in our way but to help us to comply with one common set of rules, a "total harmonization" that builds consumer confidence in what we do.

✈   European and international standards (EN and IEC) are fast becoming the de facto worldwide requirements for product safety, EMC, quality, and the environment.

The manufacturer is responsible for designing equipment in accordance with the essential requirements and the harmonized standards. You should have a general understanding of the directives for enforcement and procedural issues, but *focus on the standards* for product design and conformity assessment. When a product conforms to the standards we achieve a "presumption of conformity" to the essential requirements of the directives. The end-product manufacturer is accountable for *all* conformity aspects for their products, including the design, components, documentation, declaration, and CE marking. If the equipment becomes suspect, or there is an incident, shifting the blame to others (such as a consultant, test lab, or component supplier) is not possible. Making sound decisions, such as sourcing EU type–approved components and obtaining certification on the end-product may not at first seem to be the most convenient path, but in the end it is the easiest and lowest-cost alternative. Putting consumer safety first helps to limit the suppliers risks as well as to achieve your main goal of protecting the customer (consumer/user/operator). Figure 7-2 shows the steps to European conformity.

Let's review a few important European Conformity definitions:

- *Directives.* Directives are European laws that detail the legal, procedural, and CE marking requirements for products. Directives are published in the *Official Journal* and describe in "general" terms the essential health and safety requirements that must be met to allow products to be "placed on the market" and ensure the "free movement of goods." The directives also mandate the publication of safety and EMC standards that equipment manufacturers must follow to obtain the "presumption of conformity."
- *Standards.* European harmonized standards provide "specific" technical safety/EMC design, testing and pass/fail rules for components, products, and machines. Standards are the cornerstone of the New Approach, and without them the Single Market could not function. The EU standards, referred to in the directives, provide a "presumption of conformity" for equipment manufactured in accordance with these standards. Without properly applying EU standards, equipment will not benefit from a presumption of conformity conferred by the use of such standards.
- *CE marking.* CE is the equipment manufacturer's self-declaration symbol to indicate the equipment's conformity with all relevant directives. In addition to the CE symbol, the manufacturer's declaration of conformity and technical file must be readily available. The CE marking and declaration together are a passport for trade that indicates conformity to customs inspectors and authorities allowing products to be "placed on the market" and that ensures the "free movement of goods." They also give enforcement authorities the ability to "withdraw non-conforming products" and to take appropriate measures. CE is applied to complete products and is not intended for most components. The CE marking is not a safety/EMC guarantee from a third party (notified body) or quality mark and it is not intended for sales or marketing purposes.

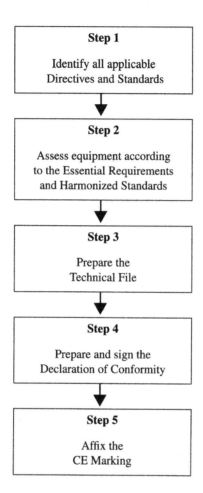

Step 1

Identify all applicable
Directives and Standards

Step 2

Assess equipment according
to the Essential Requirements
and Harmonized Standards

Step 3

Prepare the
Technical File

Step 4

Prepare and sign the
Declaration of Conformity

Step 5

Affix the
CE Marking

Note: All five steps are required, whether the manufacturers are completing the process on their own for self-declaration or using an accredited EU body for the voluntary or mandatory certification procedures. For certification, the EU testing and certification body will perform steps one and two, and oftentimes step three.

FIGURE 7-2 The Steps to European Conformity (CE).

- *Technical file.* The manufacturer's technical documentation file supports the CE marking. The contents of a technical file generally include the declaration of conformity, name and address of manufacturer, product description and identification, list of standards and directives applied, design and schematic drawings, calculations, test reports, parts lists, manuals, and so on. Technical files must be readily available and may be requested by enforcement authorities for inspection purposes should a product become suspect or an incident occur.

- *Test report.* Test reports are technical records on the conformity assessment of a product according to specific standards. Test reports are concise accounts that include clause-by-clause details on the results of a product assessment, standards rationale, test data, safety/EMC construction, and critical components. Test reports are an essential tool for conformity assessment of equipment and the key element of a technical file. Test reports contain the results of the conformity assessment and need not contain confidential design information. When a EU approval mark is not affixed to the component, product, or machine in question, test reports may be requested by customers, testing bodies, or enforcement authorities for review and verification purposes. In the event that conformity is challenged, the submission to the enforcement authority of a report drawn up by a notified body (i.e., VDE/TUV) is considered an element of proof and evidence that the equipment complies with the safety/EMC objectives.
- *Certificates and approval-marks.* Visible attestations from European accredited testing and certification bodies that provide positive evidence of conformity. Approval marks are product quality marks for safety/EMC compliance. Certificates and marks are supported by verifiable and accurate test reports from EU-notified or competent bodies. Marks and certificates establish the independent verification of a product's conformity that customers may demand. Regular follow-up inspections are performed, on equipment with EU approval marks, by the certification body to ensure ongoing conformity. When issued by recognized European certification bodies, official certificates and test reports provide the ultimate "defense of due diligence" should a product's conformity come into question.

Since the meaning of *safety* varies between individuals and depends on our point of view, I have purposely left the definition of safety till the end. With a thorough knowledge of safety, we can now answer the question that has been the topic of many discussions, "What is a safe product?" The General Product Safety Directive defines a "safe product" as "any product which, under normal or reasonably foreseeable conditions of use does not present any risk or only minimal risks compatible with the products use, considered as acceptable and consistent with a high level of protection for the safety and health of persons." Also, as stated in the Product Liability Directive, a "defective product" is one that "does not provide the safety which a person is entitled to expect, taking all circumstances into account."

Along with the formal interpretation of safety mentioned above my definition should also be considered: *Safety means putting consumer protection first and manufacturing products that conform to harmonized standards as the minimum criterion.*

In other words, meeting all the relevant harmonized standards is the *minimum acceptable level,* and the manufacturer may actually have to go beyond them to (1) satisfy the "essential requirements" of the directives, (2) meet the "latest state of technology," and (3) achieve the "high level of safety" the consumer requires.

A "Test Report and Approval Mark" from a highly reputable European testing and certification body builds consumer confidence in the equipment's safety, EMC, and quality. To adequately protect the consumer and manufacturer, this safety/EMC certification and approval marking philosophy should be applied to all equipment categories, including components, products, and machinery. Manufacturer's whose equipment truly meet the goals of European Conformity will pass the final test as "marketable and approved!"

Opinions are free of charge, but facts take charge.

DAVID LOHBECK

Appendix A: European Safety and EMC Acronyms

AIMD	Active Implantable Medical Devices Directive
BG	Berufsgenossenschaften (Trade Cooperative Association)
CB	Certification Body
CCA	CENELEC Certification Agreement
CCB	Committee of Certification Bodies
CD	Committee Draft
CE	Conformite Europeene (European Conformity)—"CE Marking"
CEN	European Committee for Standardization
CENELEC	European Committee for Electrotechnical Standardization
CERT	EC Certificate of Conformity (by Notified Body)
DIS	Draft International Standard
DOA	Date of Announcement (EN standards)
DOP	Date of Publication (EN standards)
DOR	Date of Ratification/Implementation (EN standards)
DOW	Date of Withdrawal (EN standards)
E	Draft (Standard)
EC	European Community (see EU)
ECB	European Competent Body (for EMC)
EEA	European Economic Area
EEC	European Economic Community
EFTA	European Free Trade Association
EHSR	Essential Health & Safety Requirements
EMC	Electromagnetic Compatibility
EMCD	EMC Directive
EMD	European Machinery Directive
prEN	Draft European Standard (preliminary)
EN	European Norm (European Standard)
prENV	Draft European Prestandard
ENV	European Prestandard (temporary standard)
EOTC	European Organization for Testing and Certification

ER's	Essential Requirements (see EHSR)
EU	European Union
EUT	Equipment under Test (for EMC)
FCS	Full Certification Scheme
GS	Geprufte Sicherheit (safety tested)—German safety mark
GSG	Geratesicherheitsgesetz (Equipment Safety Law)
HD	Harmonization Document (similar to a European Standard)
IEC	International Electrotechnical Commission
ISM	Industrial, Scientific, and Medical
ISO	International Organization for Standardization
ITE	Information Technology Equipment
IVD	In Vitro Diagnostic Devices Directive
LVD	Low Voltage Directive (Product Safety Law)
MDD	Medical Devices Directive
MOU	Memorandum of Understanding
MRA	Mutual Recognition Agreement
NCB	National Certification Body
OJ	*Official Journal* (see *OJEC*)
OJEC	*Official Journal of the European Communities*
PTT	Post, Telephone, and Telegraph (for telecom)
SMT	Supervised Manufacturers Test Scheme
TBM	Testing by Manufacturer Scheme
TC	Technical Committee
TCF	Technical Construction File
TMP	Test at Manufacturer's Premises
TRF	Test Report Format (officially recognized EU format)
TUV	Technischer Uberwachungs–Verein (Technical Surveillance Association)
VDE	Verband Deutscher Elektrotechniker (Association of German Electrical Engineers)

Appendix B: Information Sources

— **Standards** —

American National Standards
  Institute (ANSI)
  11 West 42nd Street
  New York, NY 10036
  tel: 212-642-4900
  fax: 212-302-1286

British Standards Institute (BSI)
  PO Box 16206
  Chiswick,
  London, W4 4ZL, UK
  tel: 011-44-181-996-7000
  fax: 011-44-181-996-7001

Global Engineering Documents
  15 Inverness Way East
  Englewood, CO 80112
  tel: 800-854-7179
  fax: 303-397-2740

International Electrotechnical
  Commission (IEC)
  3, rue de Varembé, PO Box 131
  1211 Geneva 20, Switzerland
  tel: 011-41-22-919-0228
  fax: 011-41-22-919-0300

International Organization for
  Standardization (ISO)
  Case Postale 56
  1211 Geneva 20, Switzerland
  tel: 011-41-22-7490336
  fax: 011-41-22-7341079

VDE–Verlag GmbH
  Bismarkstrasse 33
  10625 Berlin, Germany
  tel: 011-49-30-348001-0
  fax: 011-49-30-3417093

— **Directives** —

Office of the European Union and
  Regulatory Affairs
  US Dept of Commerce—Room 3036
  14th & Constitution Ave, N.W.
  Washington, D.C. 20230
  tel: 202-482-5276
  fax: 202-482-2155

Office of the Official Publications
  of the EC
  2 rue Mercier
  L2144 Luxembourg
  tel: 011-352-29291
  fax: 011-352-292942763

— European Commission Guidelines —

- Guide to the Implementation of Community Harmonization Directives Based on the New Approach and Global Approach
- Guidelines on the Application of Low-Voltage Directive 73/23/EEC Amended by the Directive 93/68/EEC
- Guidelines on the Application of Council Directive 89/336/EEC of 3 May 1989 on the Approximation of the Laws of the Member States Relating to Electromagnetic Compatibility
- Community Legislation on Machinery: Comments on Directive 89/392/EEC and Directive 91/368/EEC
- Machinery: Useful Facts in Relation to Directive 89/392/EEC

— Newsletters and Information —

SAFETY and EMC — ERA Technology Ltd
Cleeve Road, Leatherhead, Surrey
KT22 7SA, UK
tel: 011-44-1372-367014
fax: 011-44-1372-377927
e-mail: pub.sales@era.co.uk

APPROVAL — M&M Business Comm.
Lime Tree House, Lime Tree Walk,
Sevenoaks, Kent, TN13 1YH, UK
tel: 011-44-1732-746616
fax: 011-44-1732-746617
e-mail: cemark@approval.cix.co.uk

VDE Certification — VDE Institute
Merianstrasse 28
D–63069 Offenbach, Germany
tel: 011-49-69-8306-0 (Institute), or
011-49-69-8306-224 (publication)
fax: 011-49-69-8306-555, or -777

EUROPE-LINK — Single Market
Ventures
87 Rue Faider
B–1050, Brussels, Belgium
tel: 011-322-537-2603
fax: 011-322-537-1078
e-mail: europe.link@skynet.be

Index

Printed and bound by CPI Group (UK) Ltd, Croydon, CR0 4YY

03/10/2024

01040433-0009